Experiments with Plants

TEACHER'S GUIDE

SCIENCE AND TECHNOLOGY FOR CHILDREN

NATIONAL SCIENCE RESOURCES CENTER
Smithsonian Institution–National Academy of Sciences
Arts and Industries Building, Room 1201
Washington, D.C. 20560

NSRC

The National Science Resources Center is operated by the National Academy of Sciences and the Smithsonian Institution to improve the teaching of science in the nation's schools. The NSRC collects and disseminates information about exemplary teaching resources, develops and disseminates curriculum materials, and sponsors outreach activities, specifically in the areas of leadership development and technical assistance, to help school districts develop and sustain hands-on science programs. The NSRC is located in the Arts and Industries Building of the Smithsonian Institution and in the Capital Gallery Building in Washington, D.C.

National Science Resources Center

Douglas Lapp, Executive Director
Charles N. Hardy, Deputy Director for Information
    Dissemination, Materials Development, and Publications
Sally Goetz Shuler, Deputy Director for Development,
    External Relations, and Outreach
R. Gail Thomas, Administrative Officer
Anne E. Pomerleau, Financial Associate
Gail Greenberg, Executive Administrative Assistant
Heather C. Schofield, Administrative Assistant
Tonya Miller, Receptionist/Office Assistant
Kathleen Holmay, Public Information Consultant
Richard M. Witter, Development Consultant

Publications

Dean Trackman, Director
Linda Harteker, Writer/Editor
Lynn Miller, Writer/Editor
Dorothy Sawicki, Writer/Editor
Max-Karl Winkler, Illustrator
Heidi M. Kupke, Publications Technology Specialist
Matthew Smith, Editorial Assistant
Laura Akgulian, Writer/Editor Consultant
Cynthia Allen, Writer/Editor Consultant
Judith Grumstrup-Scott, Writer/Editor Consultant
Lois Sloan, Illustrator Consultant

Science and Technology for Children Project

Judith White, Program Officer, STC Discovery Deck
Wendy Binder, Research Associate
Edward Lee, Research Associate
Christopher Lyon, Research Associate
Carol O'Donnell, Research Associate
Lisa Bevell, Program Assistant
Amanda Revere, Program Aide

Outreach

Linda J. Bentley, Outreach Coordinator
Leslie J. Benton, Program Officer, Technical Assistance
Julie Clyman Lee, Program Associate
Cathy Gruber, Program Assistant
Megan Thomas, Receptionist/Office Assistant

Information Dissemination

Evelyn M. Ernst, Director
Marilyn Fenichel, Program Officer
Rita C. Warpeha, Resource/Database Specialist
Barbara K. Johnson, Research Associate
Sharon S. Seaward, Program Assistant

The above individuals were members of the NSRC staff in 1996.

ISBN 0-89278-680-9

Published by Carolina Biological Supply Company, 2700 York Road, Burlington, NC 27215.
Call toll free 800-334-5551.

See specific instructions in lessons and appendices for photocopying.

♻ Printed on recycled paper.

Foreword

Since 1988, the National Science Resources Center (NSRC) has been developing Science and Technology for Children (STC), an innovative hands-on science program for children in grades one through six. The 24 units of the STC program, four for each grade level, are designed to provide all students with stimulating experiences in the life, earth, and physical sciences and technology while simultaneously developing their critical thinking and problem-solving skills.

Sequence of STC Units

Grade	Life, Earth, and Physical Sciences			
1	Organisms	Weather	Solids and Liquids	Comparing and Measuring
2	The Life Cycle of Butterflies	Soils	Changes	Balancing and Weighing
3	Plant Growth and Development	Rocks and Minerals	Chemical Tests	Sound
4	Animal Studies	Land and Water	Food Chemistry	Electric Circuits
5	Microworlds	Ecosystems	Motion and Design	Floating and Sinking
6	Experiments with Plants	Measuring Time	The Technology of Paper	Magnets and Motors

The STC units provide children with the opportunity to learn age-appropriate concepts and skills and to acquire scientific attitudes and habits of mind. In the primary grades, children begin their study of science by observing, measuring, and identifying properties. Then they move on through a progression of experiences that culminate in grade six with the design of controlled experiments.

Sequence of Development of Scientific Reasoning Skills

Scientific Reasoning Skills	Grades					
	1	2	3	4	5	6
Observing, Measuring, and Identifying Properties	♦	♦	♦	♦	♦	♦
Seeking Evidence Recognizing Patterns and Cycles		♦	♦	♦	♦	♦
Identifying Cause and Effect Extending the Senses				♦	♦	♦
Designing and Conducting Controlled Experiments						♦

The "Focus–Explore–Reflect–Apply" learning cycle incorporated into the STC units is based on research findings about children's learning. These findings indicate that knowledge is actively constructed by each learner and that children learn science best in a hands-on experimental environment where they can make their own discoveries. The steps of the learning cycle are as follows:

- Focus: Explore and clarify the ideas that children already have about the topic.

- Explore: Enable children to engage in hands-on explorations of the objects, organisms, and science phenomena to be investigated.

- Reflect: Encourage children to discuss their observations and to reconcile their ideas.

- Apply: Help children discuss and apply their new ideas in new situations.

The learning cycle in STC units gives students opportunities to develop increased understanding of important scientific concepts and to develop better attitudes toward science.

The STC units provide teachers with a variety of strategies with which to assess student learning. The STC units also offer teachers opportunities to link the teaching of science with the development of skills in mathematics, language arts, and social studies. In addition, the STC units encourage the use of cooperative learning to help students develop the valuable skill of working together.

In the extensive research and development process used with all STC units, scientists and educators, including experienced elementary school teachers, act as consultants to teacher-developers, who research, trial teach, and write the units. The process begins with the developer researching the unit's content and pedagogy. Then, before writing the unit, the developer trial teaches lessons in public school classrooms in the metropolitan Washington, D.C., area. Once a unit is written, the NSRC evaluates its effectiveness with children by nationally field-testing it in ethnically diverse urban, rural, and suburban public schools. At the field-testing stage, the assessment sections in each unit are also evaluated by the Program Evaluation and Research Group of Lesley College, located in Cambridge, Mass. The final editions of the units reflect the incorporation of teacher and student field-test feedback and of comments on accuracy and soundness from the leading scientists and science educators who serve on the STC Advisory Panel.

Major support for the STC project has been provided by the National Science Foundation, the John D. and Catherine T. MacArthur Foundation, the U.S. Department of Defense, the Dow Chemical Company Foundation, and the U.S. Department of Education. Other contributors include E. I. du Pont de Nemours & Company, the Amoco Foundation, Inc., and the Hewlett-Packard Company.

Acknowledgments

The primary author of the *Experiments with Plants* unit was Patricia McGlashan. In developing the unit, she worked closely with Paul Williams, professor of plant pathology at the University of Wisconsin, who developed the Wisconsin Fast Plants™, and with Coe Williams and Jane Sharer, Coordinators of the Wisconsin Fast Plants Project. John M. Greenler, Staff Scientist and Co-coordinator of the Bottle Biology Project and the Wisconsin Fast Plants™ Program, served as one of the scientific reviewers of the unit. Pat also worked with a number of individuals, including:

Richard Efthim, Manager, Naturalist Center, Smithsonian Institution

Richard Hofmeister, Chief, Special Assignments and Photography Branch, Smithsonian Institution, Washington, D.C.

Sally Love, Director, Insect Zoo, Smithsonian Institution

Carolyn Margolis, Project Coordinator, Quincentenary Program, Smithsonian Institution

Joan Nowicke, Curator, Department of Botany, Smithsonian Institution

Dane Penland, Staff Photographer, Smithsonian Institution Photographic Services

Barbara van Creveld, Program Manager, Insect Zoo, Smithsonian Institution

Staff of the Entomology and Botany Libraries of the Natural History Libraries, Smithsonian Institution

Staff of the Office of Horticulture, Smithsonian Institution

Many people contributed to the national field testing of the *Experiments with Plants* unit. The following individuals coordinated the testing:

Marcia J. Massey, Curriculum Specialist, Hartford Public Schools, Hartford, CT

Shelley Barbary, Science Consultant, School District of Greenville County, Greenville, SC

Jack Greene, Fairfax County Public Schools, Annandale, VA

Neal Eigenfeld, Milwaukee Public Schools, Milwaukee, WI

Allen Stawicki, Morse Middle School, Milwaukee, WI

The NSRC would like to thank George Hein and Sabra Price of Lesley College for evaluating this unit. The NSRC also would like to thank the following individuals and school systems for their assistance with the national field testing of the *Experiments with Plants* unit:

In the Hartford Public Schools:
Juanita Beasley, Martin Luther King School;
James Ciancimino, Martin Luther King School;
Barbara DeMaio, Martin Luther King School;
Shirley Paddyfoote, Martin Luther King School;
Joseph V. Pandolfio, C. A. Barbour Elementary School

In the School District of Greenville County:
David Ammons, Morton Elementary School;
Tina Dillard, Duncan Chapel Elementary School; Garrison Hall, Taylors Elementary School

In the Fairfax County Public Schools:
Betty Eubanks, Fairview Elementary School

In the Milwaukee Public Schools:
Erin Benedetto, Walker Middle School; Kelly Clark, Walker Middle School; Merry Beth Gerschke, Morse Middle School

STC Advisory Panel

Contents

Goals for *Experiments with Plants*

In this unit, students plan and conduct experiments to determine how different variables affect the growth and seed production of rapid-cycling *Brassica rapa* (Wisconsin Fast Plants™). Their experiences introduce them to the following science concepts, skills, and attitudes.

Concepts

- Plants need soil nutrients, light, and water.

- Plant growth is affected by the quantities of nutrients, light, and water available.

- Controlling variables enables the effect of each to be identified and studied.

- Flowering plants must be pollinated in order to produce seeds.

- Bees are effective pollinators.

- One seed has the potential to produce one plant.

- The number of seeds produced by a single plant is affected by such variables as nutrients, light, water, and the extent of pollination.

- The orientation of a plant's growth is affected by gravity and light.

Skills

- Planting and caring for plants.

- Predicting how changing one variable might affect the outcome of an experiment.

- Planning and conducting experiments in which variables are controlled.

- Observing, measuring, describing, and recording changes in plant growth.

- Communicating results through graphs, drawings, and group presentations.

- Interpreting and analyzing how different variables affect the growth and change of plants over time.

- Reflecting on experiences through writing and discussion.

- Reading and researching to learn more about plants.

Attitudes

- Developing an interest in investigating plant growth.

- Appreciating the need for careful and precise design of experiments.

- Appreciating the need for detailed recordkeeping during experimentation.

- Valuing scientific data that has been collected over time.

Unit Overview and Materials List

Experiments with Plants is an 8-week, 16-lesson unit designed and tested for 5th and 6th graders that features rapid-cycling Wisconsin Fast Plants™ as a vehicle for experimentation. These plants are *Brassicas*, members of the mustard and cabbage family, and they were developed by Paul Williams of the University of Wisconsin. Wisconsin Fast Plants are especially well-suited for classroom study because they are hardy, compact, thrive under artificial lights, and complete their life cycle in about 40 days.

This unit is a sequel to the STC project's 3rd grade unit, *Plant Growth and Development*. Some prior study of plants and plant life cycles will be helpful to students, as offered by the *Plant Growth and Development* unit, but is not essential. The timetable on pg. 4 will help you correlate the life cycle of these plants with the lessons of the unit.

The main objective of *Experiments with Plants* is to teach students how to design and conduct controlled investigative experiments. In the first phase of the unit, students learn through discussion and reading to identify the key variables that affect the life, health, and reproductive capabilities of Wisconsin Fast Plants, and they learn they can manipulate these variables. Then, working in teams, students formulate a question they would like to attempt to answer through an experiment involving one of the variables, and they proceed to design and set up team experimental plans.

In the second and longest phase—about 6 weeks—students execute their team experiments and collect data. They plant Wisconsin Fast Plants seeds according to their team's experimental plans, and the plants begin to grow. From that point forward through Lesson 9, students collect data on a daily basis. Through data collection, measurement, observation, and recording, students discover the effects on the plants of their manipulation of their chosen variable. Students most closely observe the effects of their own experiment (the lack of growth in a plant deprived of fertilizer, for example) but also get the opportunity to discuss with other students the effects of their experiments.

Management of the unit can present somewhat of a challenge throughout the second phase, so special instructions are provided in the Teacher's Guide. These special instructions are flagged with icons. The appearance of the icon of the finger tied with a bow indicates special instructions, primarily regarding equipment and the need to remember to set it up in advance of student use. (See **Teaching Strategies and Classroom Management Tips** section, pg. 5.)

After fertilization of the plants by cross-pollination, during which students learn about the interdependent relationship of bees and flowering plants, the plants develop seed pods. Then the mature plants are allowed to die as their seed pods dry and ripen, and students count the seeds in a final collection of data.

In the third phase, with the life cycle of the plant complete, students reflect on their experiences by reviewing their journals of data and observations. They compare and contrast their own data with the data from other team members. Then the teams attempt to draw conclusions and decide if they have answered their experimental question.

In this phase, students also prepare and communicate their findings. With the teacher, they decide whether their presentations should be simple oral reports or elaborate multi-media conferences. Here students get an opportunity to interact the way scientists interact and to bring the major experiment to an enjoyable conclusion that, depending on the form of presentation chosen, can involve others in the school and their parents.

At the end of the unit are two sets of experiments involving germination and tropisms. The germination experiments (Lessons 12 and 13) use seeds harvested from the first set of plants, helping students see the continuous nature of the life cycle. The germination as well as the tropism experiments (Lessons 14 and 15) also provide excellent opportunities for reinforcement of the skills and knowledge acquired earlier in the unit and could serve as useful evaluations of student learning. Additional post-unit assessments are provided in **Appendix A**.

You do not have to be an expert in botany to teach this unit. The background sections of the Teacher's Guide will provide you with most of the information you need. But don't be surprised if you find yourself learning along with the students, and if you and your students encounter some puzzling questions. Use this situation to model the way scientists learn: define the question, then ask, "How can we find out?" This will encourage your students to find their own answers by experimenting and consulting resource materials.

Materials List

Below is a list of materials for the *Experiments with Plants* unit.

1	Teacher's Guide
15	Student Activity Books
30	trays
1 pkg.	toothpicks
30	spoons
15	droppers
30	small paper cups
60	small plastic cups
6	large paper cups
1 pkg.	dried honeybees
15	pairs of forceps
15	hand lenses
2 pkgs.	Wisconsin Fast Plants™ seeds
2 pkgs.	fertilizer
2 pkgs.	potting mix
36	planter labels
34	planter quads
3	water tanks
3	water mats
144	wicks
3	felt squares containing copper sulfate (see pg. 6)
3 pkgs.	wooden stakes
3 pkgs.	plastic rings
1	lighting system

1	extension cord
60	small resealable plastic bags
30	sheets of transparency film
3	thermometers
*	Student notebooks
*	Drawing paper
*	Blank index cards, cut to 1½" x 5"
* 1	large jar
*	24" x 36" newsprint pad and markers OR
*	Overhead transparencies and markers, with projector and screen
*	Paper towels
*	Sponges
* 2	dish pans
*	Whisk broom and dustpan
*30	pairs of scissors
*	Glue and small glue cups
*	Crayons
*	Staplers

*Note: These items are not included in the kit. Including them would increase material and shipping costs, and they are commonly available in most schools or can be brought from home.

Timetable for *Experiments with Plants*

Lesson No.	Activity	Date												
	Preparation													
	Assemble lights													
	Set up watering system													
1, 2, 3	Plan experiments													
4	Plant the seed (on a Monday or Tuesday)	0												
	Water from top	0 1 2												
5	Thin and transplant		4 5											
	Check water level		4											
	Begin graphing plant's height		4 5											
	Observe true leaves and flower buds develop		6 7 8											
	Observe growth spurt			7 8 9 10 11 12 13 14										
6	Make bee stick			9 10										
	Check water level			11										
	Observe flowers open			11 12 13 14 15 16 17										
7	Pollinate				12 13 14 15 16 17 18									
	Pinch off unopened buds				18									
	Check water level				18									
	Observe seed pods develop				16 17 18 19 20 21 22 23 24 25 26 27 28 29 30 31 32 33 34									
	Check water level					25								
	Remove plants from water					35								
8	Harvest and thresh seeds						42							
	Clean up and store equipment													
12, 13	Germination experiments													
14, 15	Tropism experiments													

Key: ■ Do activity on this day

☐ Do activity on one of these days

*This timetable as it relates to the Wisconsin Fast Plants™ life cycle is for the control plants. Expect the life cycle of experimental plants to be different from the control plants.

Teaching Strategies and Classroom Management Tips

The teaching strategies and classroom management tips in this section will help you give students the guidance they need to make the most of the hands-on experiences in this unit. These strategies and tips are based on the assumption that students already have formed many ideas about how the world works and that useful learning results when they have the opportunity to re-evaluate their ideas as they engage in new experiences and encounter the ideas of others.

Classroom Discussion: Class discussions, effectively led by the teacher, are important vehicles for science learning. Research shows that the way questions are asked, as well as the time allowed for responses, can contribute to the quality of the discussion.

When you ask questions, think about what you want to achieve in the ensuing discussion. For example, open-ended questions, for which there is no one right answer, will encourage students to give creative and thoughtful answers. Other types of questions can be used to encourage students to see specific relationships and contrasts or to help them to summarize and draw conclusions. It is good practice to mix these questions. It also is good practice always to give students "wait-time" to answer. (Some researchers recommend a minimum of 3 seconds.) This will encourage broader participation and more thoughtful answers.

Brainstorming: A brainstorming session is a whole-class exercise in which students contribute their thoughts about a particular idea or problem. It can be a stimulating and productive exercise when used to introduce a new science topic. It is also a useful and efficient way for the teacher to find out what students know and think about a topic. As students learn the rules for brainstorming, they

will become more and more adept in their participation.

To begin a session, define for students the topics about which ideas will be shared. Tell students the following rules:

- Accept all ideas without judgment.

- Do not criticize or make unnecessary comments about the contributions of others.

- Try to connect your ideas to the ideas of others.

Ways to Group Students: One of the best ways to teach hands-on science lessons is to arrange students in small groups of two to four. There are several advantages to this organization. It offers pupils a chance to learn from one another by sharing ideas, discoveries, and skills, and, with coaching, students can develop important interpersonal skills that will serve them well in all aspects of life. Finally, by having children help each other in groups, you will have more time to work with those students who need the most help.

As students work, often it will be productive for them to talk about what they are doing, resulting in a steady hum of conversation. If you or others in the school are accustomed to a quiet room, this new, busy atmosphere may require some adjustment. It will be important, of course, to establish some limits to keep the noise under control.

Among the activities in this unit that students will be working together on are: planning the team experiment; comparing and contrasting data; and presenting team conclusions.

Safety: This unit contains nothing of a highly toxic nature, but common sense dictates that nothing be put in the mouth. In fact, it is good practice to tell your students that, in science, materials are never tasted. Students may also need to be reminded that certain items, such as toothpicks, forceps, and water droppers, are not toys and should be used only as directed.

The anti-algal substance in the felt squares is copper sulfate, which dissolves in the water to prevent algal growth. As a solid, copper sulfate can be irritating to the skin; also small amounts of the solid can be toxic and must not be ingested. However, the amount of copper sulfate used in the felt squares is very small and very dilute. Finally, like all flowering plants, Wisconsin Fast Plants™ produce pollen, which can be an allergen.

Handling Materials: To help ensure an orderly progression through the unit, you will need to plan ahead and to establish a system for storing and distributing materials. Being prepared is the key to success. Here are a few suggestions.

- Check your Timetable for *Experiments with Plants* daily (see pg. 4). Know which activity is scheduled and which materials are going to be used.

- Familiarize yourself with the materials as soon as possible. It might be useful to label everything and spread it out on a table for students to see.

- Organize your students so that they are involved in distributing and returning materials. If you have an existing network of cooperative groups, delegate the responsibility to one member of each group.

- Organize a distribution station and train your students to pick up and return supplies to that area. Depending on the activity, this might be as simple as picking up a seed and a hand lens or as complicated as assembling all of the supplies needed to plant the seeds. A cafeteria-style approach works especially well when there are large numbers of items to distribute.

- Preview each lesson ahead of time. Some have specific suggestions for handling materials needed that day.

- Familiarize yourself with the lighting system; it is the most important component of the life support system for Wisconsin Fast Plants, as they have been selectively bred to grow under continuous

(24 hours a day) cool white fluorescent light. Once the seeds have been planted, the lights must go on and stay on (you probably will want to leave a note for your custodial staff). Be sure to keep the lights above the plants, by about 5 to 7.5 cm (2 to 3 inches); additional instructions are contained in the kit.

Additional management tips are provided throughout the unit. Look for following icon.

Refer to the **Unit Overview** and **Materials List** for more materials information.

Setting up a Learning Center: Supplemental science materials should be given a permanent home in the classroom in a spot designated as the learning center. Such a center could be used by students in a number of ways: as an "on your own" project center, as an observation post, as a trade book reading nook, or simply as a place to spend unscheduled time when assignments are done.

In order to keep interest in the center high, change it or add to it often. Here are a few suggestions of items to include:

- science trade books on plants, insects, and famous scientists

- snap-together centimeter cubes, a centimeter ruler, and a balance scale with an assortment of interesting objects to measure and weigh, such as paper clips, a box of cereal, wooden blocks, a lunch box, vegetables, and canned goods

- magnifying glasses and an assortment of interesting objects to observe, such as leaves, seeds, stems, roots, flowers, insects, soil and rocks, newspaper, fabric scraps, salt, feathers

- calculators

- audiovisual materials on related subjects, such as plants, insects, interdependence, pollination, life cycles, or famous scientists and famous experiments

- other live plants, especially those started from seed that grow at a normal rate

- models of bees and brassica plants

- items contributed by students for sharing, such as an insect collection, a honeycomb, magazine or newspaper articles featuring graphs, pictures, books, beeswax candles, pressed plants, seed collections, pods, and model farm equipment

Curriculum Integration: There are many opportunities for curriculum integration in this unit. Look for the following icons for math, reading, writing, art, and speaking that highlight these opportunities.

Evaluation: In the STC project, evaluation tools are included throughout each unit, and post-unit assessments are provided at the end. This arrangement is intended to help you assess what students know and monitor how they are progressing, making it easier for you to provide assistance to students who need it, to go over materials students did not grasp, and to report to parents on student progress. The assessments provided also are intended to be directly helpful to students, giving them an opportunity to reflect on their own learning, gain confidence by viewing their own progress, articulate the ways in which they want and need to grow, and formulate further questions.

Evaluation Preparation: To facilitate successful documentation and assessment of students' learning in the specific content areas of the unit as well as in the development of relevant skills, you should be prepared to:

- **Collect pre/post information.** One of the best indicators of student learning comes through gathering information from an identical activity or discussion conducted both before and after a unit. The pretest suggested for use in Lesson 1 of this unit also is a suggested post-test. Note that comparison of non-identical materials generated early and late in a unit also can help you gauge growth in learning. You will want to make sure that pretest and post-test work is dated.

- **Encourage students to use notebooks in a way that is useful to you and to them.** Student notebooks provide information about student progress. To ensure that

you will be able to make the best use of these notebooks, be sure to ask students to:

- make only one entry per page;
- date each entry;
- write out conclusions and interpretations of experiments;
- write explanations to charts, tables, and graphs; and
- include the question when writing out answers.

Student notebooks are a particularly handy and effective way to share student progress and accomplishments with parents and other interested adults as well as with the students themselves.

- **Save other student work.** Other student work, such as pollution prevention barriers, lightproof boxes, and graphic presentations of the project, provide concrete demonstrations of student knowledge.

- **Observe.** Invaluable for assessment are your ongoing observations of students as they work, written in your own notebook or on file cards.

- **Allow time for oral presentations.** Oral presentations by students can be useful vehicles for assessment.

A variety of post-unit assessment instruments is offered in **Appendix A**. From those suggested, you will want to choose only the instruments that are most appropriate for measuring the achievement level of your class. Consider using more than one, in order to give students with differing learning styles a chance to express their knowledge and skills. Different styles of assessment have been shown to be particularly helpful in increasing the precision of the assessment of girls and minorities—two groups that have historically underperformed in science.

Please note that **Appendix A** includes a black line master for a "Teacher's Record Chart of Student Progress." You may want to reproduce this chart at the beginning of the unit to help you record individual student achievement throughout the unit. Please remember that most students at the grade level targeted by this unit will not be able to master and articulate the full list of skills.

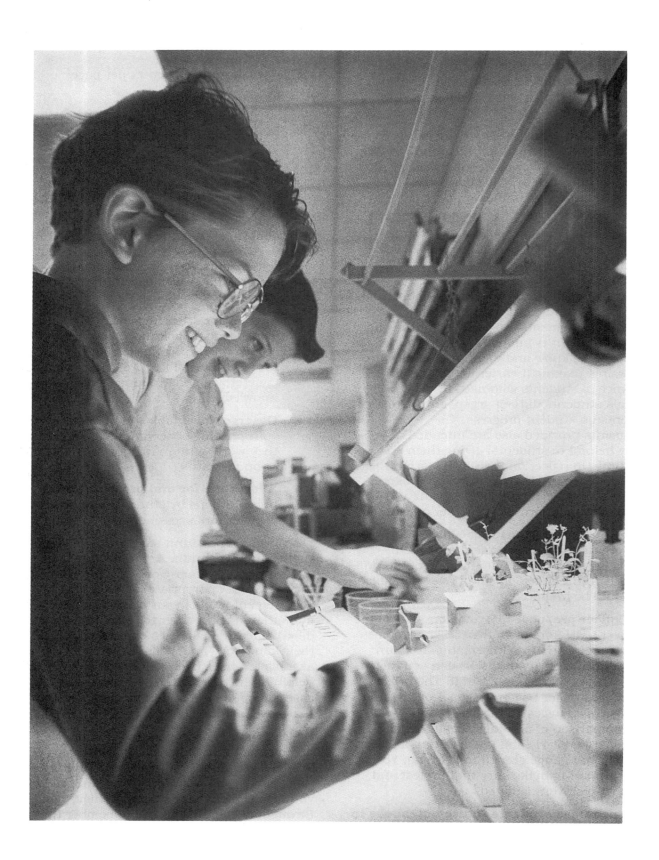

What Do You Know about Experiments?

Overview

Throughout this unit, students will be learning about scientific methodology that involves how to design **controlled science experiments**. These experiments include **variables**, conditions that change. They also include **controls**, which are conditions that remain constant. (Each of these terms is described in more detail in the **Background** section below.) After brief brainstorming sessions, students are invited to begin thinking about the ways in which a good experiment is like a **fair test**, during which only one thing is tested at a time, and everything else remains the same. Then the class is introduced to Wisconsin Fast Plants™ (also referred to as *Brassicas*), the vehicle this unit uses to develop scientific experiments. The class also learns about this plant's special requirements for growth and development.

Objectives

- The teacher determines what the students know already about the characteristics of a good experiment.
- The teacher introduces the idea that a good experiment is a "fair test."
- Students share what they know about flowering plants.
- Students read about the requirements for growth and development of Wisconsin Fast Plants.

Background

In this unit, the students will be using Wisconsin Fast Plants to design **controlled experiments**. These plants are particularly suited to this because they are specially bred to be compact, hardy, very fast growing, and prolific seed producers.

To develop experiments, the students will need first to identify the brassica plant's unique requirements for growth and development. These requirements refer to the conditions under which these plants grow best. They include the amount of light needed, the amount of water needed, the amount of fertilizer needed, and the appropriate temperature, among others. (These conditions are explained in the chart at the end of the **Background** section.)

Scientists call these conditions **variables**. During a scientific experiment, one variable is changed in some way. For example, if the brassica plant requires continuous light, changing that variable might mean that you try depriving it of light for extended periods of time. Then you would be testing the effect that light has on the plant. In order to guarantee that light and only light is being tested, all of the other variables—temperature, amount of fertilizer, amount of water—would remain unchanged.

This setup, in which one variable is changed and all the rest remain unchanged, is a **controlled experiment**. The variables that remain constant are the experiment's **controls**.

These are complex concepts, but a good way to explain them is to compare a controlled experiment to a race, which is nothing more than a "fair test." Children easily understand that, in order for a race to be fair, all participants must start from the same place and end at the same finish line. Other important variables are the time that the runner begins the race and the length and condition of the path. If all of these variables are held as constant as possible, the variable that is tested is the running ability of the racers. And, presumably, the person with the most running ability wins.

Being able to identify variables is an important skill in science. It implies that you are able to look at a situation in terms of all of its component parts, understand how each part might change, and speculate how that change might affect the whole. In some cases in this unit, students may conduct an experiment that produces no change. Then they may feel that their experiment has been a failure. In fact, that is not the case. Instead, students have found that changing the variable in that way has no great effect on the plant. This is an important, and valid, finding.

Use caution in introducing all of these new ideas, and explain them as needed when they come up. It will take the students some time to absorb them, as well as the words used for them. In the next lesson, all of these terms will appear again in context.

The components of a good experiment follow. The students probably will mention only a few of them during your initial discussions, but they will be able to list more by the end of the unit.

- The researcher develops a good plan and follows it.
- The topic chosen for the experiment is interesting and worth doing.
- The researcher gathers information on the topic either through reading or talking with experts.
- The researcher makes careful observations over a period of time.
- The researcher keeps records—accurately, honestly, and regularly.
- Something is measured.
- After collecting data, the researcher draws a conclusion based on the data.
- The findings of the experiment are communicated.

At the end of this lesson, the students will have time to read about what Wisconsin Fast Plants need to grow best. The chart on the next page summarizes that information for you. Manipulating these variables will form the basis of the students' experiments.

Wisconsin Fast Plants™

Ideal Growing Condition	Probable Consequences of Varying the Condition
Cool, white, fluorescent light	Other types of lights produce too much heat and scorch the leaves.
A continuous supply of water	Too little water leads to wilting and death.
Temperature between 70° and 80°F (21°–27°C)	Between 60°F (16°C) and 70°F (21°C), the life cycle slows down by several days. Below 60°F (16°C), most seeds fail to germinate.
*24 hours of continuous light	Any less light results in a slower life cycle, taller and skinnier plants, and fewer flowers and seeds.
*3 fertilizer pellets per cell	Underfertilizing results in a nutritional deficiency, which will produce stunted plants, pale leaf color, delayed flowering, and fewer seeds. Overfertilizing results in increased flower, pod, and leaf production; dark green leaf color; delayed flowering; and stress due to the high level of salts (fertilizer) in the soil.
*1 plant per cell	Competition for water, nutrients, light, and air results in weaker plants and a smaller yield of seeds.
*Cross-pollination	Unless blossoms are cross-pollinated, no viable seeds will be produced.

*Indicates that this is a good condition to vary for an experiment.

Materials

For each student
1   student notebook

For the class
2   large sheets of newsprint and markers
    OR
2   overhead transparencies and markers, with projector and screen

Preparation

1. Obtain the materials needed to record the students' ideas.

2. Label one sheet "What we know about how to carry out an experiment." Label the second sheet "What we know about flowering plants."

3. Ask the students to read the background information on Wisconsin Fast Plants™ entitled *Fast Plants for Fast Times* in the Student Activity Book on pgs. 4 and 5. You may want to assign it as homework. This reading selection is reproduced in this Teacher's Guide on pg. 16.

Procedure

1. Open the lesson by telling the students that they will be conducting plant experiments for the next 6 weeks using Wisconsin Fast Plants. Before beginning the experiments, they will need to do some thinking and careful planning. They also will need to do some background reading and research about plants. In order to be ready to begin recordkeeping in a few days, they will need to have student notebooks. All entries they make in their notebooks should be dated.

2. Using brainstorming techniques (see **Teaching Strategies and Classroom Management Tips**), find out what students think about how to carry out an experiment. Record all of their ideas on the sheet you have prepared for this.

3. Introduce the idea that, in order to do a proper experiment, the students need to set it up as a fair test. Get their ideas on how they would set up a fair race to learn which student in the class is the fastest runner. Which variables would stay the same? Which would change?

4. Next, find out what students already know about flowering plants by asking them to draw and label the parts of a flowering plant. Have them date the drawing and then put it away. Let them know that this is a record and that they should not go back and add to the drawing or change it.

5. Then invite the students to share verbally what they know about flowering plants. Try to elicit information on plant parts and on what they think the function of each part is. Figure 1-1 shows a plant with parts labeled. Record their ideas on the sheet you have prepared for this purpose.

Final Activities

1. If the students need more practice defining what constitutes a fair test, discuss other familiar situations to which they can relate. Examples follow.

    ■ What is the best way to study for a spelling quiz—with the television on or off? How could you find out? Design a fair test.

    ■ You spilled spaghetti on your shirt again. Mom says that the stain will come out, but only if you rinse it off with cold water immediately. There are a number of ways you could find out if mother knows best. What kinds of experiments would you suggest?

    ■ You have noticed that your dog seems to like the free sample of a new dog food, which came in the mail. Design an experiment to find out if he really prefers the new brand to his old one.

2. Because the experiments in the unit are based on Wisconsin Fast Plants and their unique requirements for growth, it is important that students become familiar with these requirements. Allow the students time to

Figure 1-1

A flowering plant

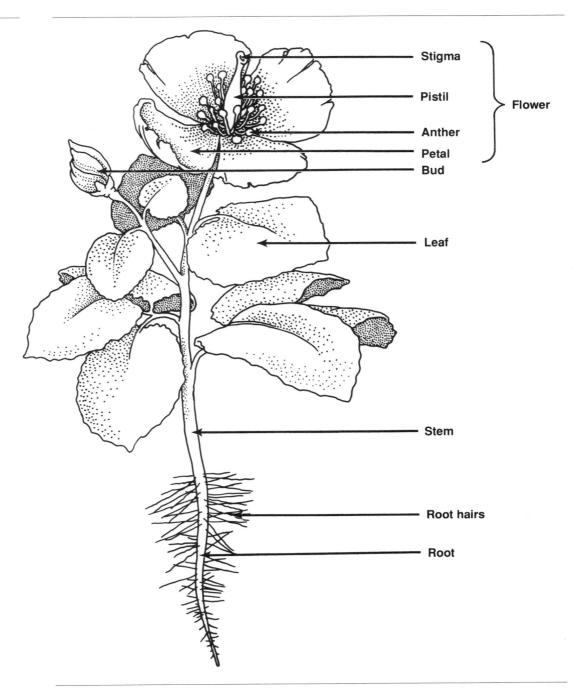

Stigma
Pistil
Anther — Flower
Petal
Bud

Leaf

Stem

Root hairs

Root

read the selection *What Do Wisconsin Fast Plants Need to Grow Best?* on pg. 6 of the Student Activity Book. The students also have been provided with their own calendar of the life cycle of the plants. The calendar is on pg. 8 of the Student Activity Book.

Extensions

1. There is tremendous variety in the plant world, as the students probably know already. Keeping this in mind, ask them to think about the following question: What would happen if they did an experiment in which they put a cactus in a pond and a water lily in the desert? What

Figure 1-2

The normal habitats for the water lily and the cactus

Wet/Temperate   Dry/Hot

changes would the students observe in each plant? The students can share their predictions with the class.

2. There are many interesting books on the subject of plants (see **Bibliography, Appendix B**). Encourage your students to do some library research. A good motivator might be a challenge to find out more about some of the following curiosities of the plant world:

■ Giant saguaro cactus (may grow to 18m [60 feet] tall and live for 200 years)

■ Rafflesia (has a flower that may weigh as much as 7 kg [15 lbs.] and stinks like rotten meat)

■ Foxglove (contains a poison that could cause a heart attack, but also is the source of digitalis, a drug used in the treatment of heart problems)

■ Coconut (the largest seed in the world, the coconut may float for months in the ocean and then sprout within a short time after being washed ashore)

■ Venus flytrap (captures insects and digests them)

■ Popcorn (pops because of water droplets trapped inside; when heated, the water becomes steam that explodes the corn)

■ Horsetails (grew 18m [60 feet] tall millions of years ago and are still around today, though smaller; because they contain abrasive silica, they also are called by another name—scouring rush)

Evaluation

The students have shared important information on what they know about how experiments are conducted and about flowering plants. They also have begun to explore the idea that a proper experiment must be a fair test of one variable. Keep your records of these opening discussions. They will give you a baseline from which to measure progress and can serve as a pre-unit assessment. Save the two sheets on which you recorded your students' ideas, and refer to them again in the post-unit assessments, when the students reflect on what they have learned about how to design and carry out an experiment.

Successful management of this unit requires that you plan ahead for each activity before it happens. Look for **Management Tips** to help you. Here is the first one.

You will want to begin setting up some of the Wisconsin Fast Plants equipment now, such as the light system and the water tanks. This will help you feel less rushed as you approach planting day. Also, the students will be able to examine the equipment, become familiar with it, and discuss how important it is to the life of the plant. Detailed instructions for setting up the lights come with the recommended lighting system (see the **Materials List**). Instructions for setting up the watering system are in Lesson 4.

**Reading
Selection**

Fast Plants for Fast Times

The Wisconsin Fast Plant™ is the plant you will be using for your experiments in this unit. It took Dr. Paul Williams, who is a professor and researcher at the University of Wisconsin, about 15 years to develop it. Fifteen years may seem like a very long time to spend breeding a plant, but think of all that he accomplished. Through selective breeding, Dr. Williams was able to speed up the plant's life cycle, making it ten times faster than that of its ancestors. Today, this small yellow-flowered plant whizzes through its entire life cycle, from seed to seed, in just 6 weeks.

Dr. Williams had an interesting reason for wanting to develop a fast plant. He is a plant pathologist, and his job is to study plant diseases and to find out if some plants inherit the ability to fight off diseases. In order to speed up his work, he needed a fast-growing plant to use in his studies.

Dr. Williams started with a world collection of more than 2,000 *Brassica* seeds and planted them in his laboratory using planting, lighting, and watering equipment almost exactly like what you will use. He observed that, out of the 2,000, only a few plants flowered much sooner than others. He took advantage of these exceptional plants by cross-breeding them. These few would be the parents of his next generation of plants. Dr. Williams wondered what kind of offspring these faster flowering parents would produce. Would the offspring inherit the ability to flower earlier than the average *Brassica* plant?

Yes! In fact, a few of the new plants even flowered a little faster than the parent plants. These slightly faster offspring were then cross-pollinated, becoming the parents of the next generation.

Dr. Williams continued to use this method of selective breeding for years. He grew populations of 288 or more plants in each generation. He cross-bred the earliest flowering plants of this population and used their seeds to grow the next generation. In each new generation, he found that about 10 percent of the plants flowered slightly earlier than their parent generation had.

The selective breeding project was a grand success. The result is what is now known as Wisconsin Fast Plants™. Besides developing a 6-week growth cycle, Dr. Williams was able to breed in other desirable qualities that make the plant a nearly ideal laboratory tool. Some outstanding traits of these plants are:

- They produce lots of pollen and eggs, resulting in many fertile seeds.
- Their seeds do not need a dormancy (or rest) period, so they can be replanted immediately.
- The plants are small and compact.
- They thrive in a crowd.
- They grow well under constant light.

Wisconsin Fast Plants have become important laboratory research tools all over the world. Soon they will be part of National Aeronautics and Space Administration's space biology program. But most exciting of all, these special plants are becoming part of school science programs across the country, from the elementary to the university level.

**Reading
Selection**

What Do Wisconsin Fast Plants Need to Grow Best?

You are about to meet a unique member of the plant world, Wisconsin Fast Plants™. *Brassica rapa* is the scientific name. For the next 6 weeks or so, you will be working very closely with these plants, observing and measuring them through all the stages of their life.

These remarkable little plants will sprout, or germinate, develop leaves and buds, flower, produce pods filled with seeds, and die—all within about 6 weeks. They are not called "fast" for nothing! To fully appreciate how quickly all of this happens, look at the calendar on pg. 19.

Because Wisconsin Fast Plants are unique, it is not surprising that they have some unusual requirements for growth and development. Let's look at each one of them in detail.

Light
The *Brassica* needs 24 hours of continuous light. Not only that, the light must come from cool, white, fluorescent bulbs.

Fertilizer
The regular dose is three fertilizer pellets per planter quad cell.

Space
One plant per planter cell is the spacing recommendation to achieve the best growth.

Pollination

Before the plants can produce seeds, pollen must be transferred from the flowers of one plant to those of another. This is known as cross-pollination.

Temperature

The plants do best when the temperature stays between 70°F and 80°F (21°C to 27°C).

Water

The plants require a continuous supply of water. They must not be allowed to dry out for more than a few hours.

These six factors—light, fertilizer, space, pollination, temperature, and water—are very important to the health and productivity of the plant. Scientists call them **variables**—conditions the plants need for growth.

Are some variables more important than others? Exactly how does each of the variables affect the plants? What will happen if you change one of the variables?

During this unit, you will be designing an experiment by changing one of these variables. Which one would you like to work with? Remember, the first step in planning a good science project is asking a good question that you can answer yourself by experimenting.

Life Cycle of a Normal *Brassica* Plant

Sunday	Monday	Tuesday	Wednesday	Thursday	Friday	Saturday
	0 DAY 0... PLANTING DAY	1	2 DAY 2 OR 3... —COTYLEDONS EMERGE —	3	4 DAYS 4, 5, OR 6... —FIRST TRUE LEAVES EMERGE —	5
6	7 ...DAYS 7, 8, OR 9.... — FLOWER BUDS APPEAR—	8	9	10 ...DAYS 10, 11, OR 12.... —GROWTH SPURT—	11	12
13	14 DAYS 14 .... — POLLINATION—	15	16 ....TO....	17	18 ......18...	19 DAYS 19.....
20 ....DAYS 19 TO 35 —SEED PODS DEVELOP—	21	22	23	24	25	26
27 ....DAYS 19 TO 35 — SEED PODS DEVELOP—	28	29	30	31	32	33
34 .... TO 35 —SEED PODS DEVELOP—	35					

The Planning Board

VARIABLE WE WILL TEST

Pollination

HOW WE WILL TEST THE VARIABLE

Don't Pollinate

VARIABLES WE WILL NOT CHANGE

Light

Water

Fertilizer

Space

Identifying Variables and Planning a Fair Test

Overview

Two ideas introduced in the first lesson are elaborated here: What are the variables that determine how well Wisconsin Fast Plants grow? What are the criteria for a fair test? By using a planning board, students can isolate the one variable they will change from those that will serve as experimental controls. The planning board is a tool that the students will use throughout the unit to understand how a controlled experiment is a fair test. In this lesson, the students will use the planning board to design their own controlled experiments with their teams.

Objectives

- Students learn more about the variables that affect plant growth.
- Students begin to learn how to conduct experiments using these variables.
- Students understand what constitutes a fair test.
- Students use planning boards to design their experiments.

Background

During this lesson, it is important to set the experimental tone of the unit. Many children have the notion that experimenting means mixing several things together to see what happens. As a departure point for younger students, this type of aimless activity can be a valuable experience. But your students are now capable of much more. In this unit, they will be imitating the method that scientists use the world over. This method is a logical process.

Planning is a crucial first step in any experiment and is vital to the success of the students' experiments. Everything depends on these next two lessons, during which the student teams design their experiments. A well-thought-out plan will result in an experiment that proceeds smoothly, provides pertinent data, and comes to a meaningful conclusion. More than that, developing a plan will give the students immediate ownership of the project, a role to play on the experimental team, a well-defined path to follow, and, in the end, a feeling of pride from accomplishment.

You will be introducing a large Planning Board in this lesson (key parts for which are provided on pgs. 136–137). The Planning Board serves as a tangible structure for you and the students to manipulate as the students plan their experiments. You will revisit the large board several times over the course of this unit to model its usefulness as a planning tool. The student version of the Planning Board is a smaller work sheet (**Activity Sheet 1A**, pg. 29) that can be snapped into the student notebook when it has been completed.

In the interest of simplicity, you should steer students toward experimenting with the four most easily managed variables: light, fertilizer, space, and pollination.

Materials

For each student
1   student notebook
1   **Activity Sheet 1A, Planning Work Sheet**

For each four-member team
1   **Activity Sheet 1B, Directions for Using the Planning Work Sheet**
1   piece of paper
1   pair of scissors

For the class
To record students' ideas:
1   large sheet of newsprint and markers
    OR
1   overhead transparency and markers, with projector and screen
1   Planning Board (see pgs. 136 to 137)
    Blank index cards for variables (cut in half to 1½" x 5")

Preparation

1.   Obtain the materials needed to record student ideas. Label the sheet "What variables do Wisconsin Fast Plants need to grow?"

2.   Make and display the Planning Board using the masters provided on pgs. 136-137. Make cards 1½" x 5" for the variables using the illustrations on pgs. 23 and 24 as examples, but do not attach the cards to the board yet.

3.   You will want to be sure that the lighting and watering systems have been set up already so that the students can see how difficult it would be to try to change one of these variables.

4.   Duplicate a copy of **Activity Sheet 1A** (pg. 29) for each student and a copy of **Activity Sheet 1B** (pg. 30) for each team.

Procedure

1.   Initiate a discussion by posing the general question, "What are all the variables that Wisconsin Fast Plants need to grow?" Record each of the variables. Your complete list might look like this:

   ■   Light

   ■   Water

   ■   Fertilizer

   ■   Space

- Pollination
- Temperature

2. Point out that each of these variables contributes to the health and productivity of the plant in some specific way. Remind the students that in Lesson 1 they read about the variables that Wisconsin Fast Plants need to grow best. Now ask students to think about and discuss how much light (water, fertilizer, space) each plant needs to grow best.

    (You may want to refer to the chart on pg. 11 in Lesson 1 to remind yourself of the ideal conditions for growing the plants and the probable consequences of changing those conditions.)

3. Mention that each of these variables will play a part in the students' experiments. Tell the class that they will be grouped into research teams of four. Each team will have four complete quads (or 16 plants) with which to work. Two of the quads should be designated the **control plants,** and two should be designated the **experimental plants**.

    Note: The control plants will be grown under the ideal conditions and will represent the normal stages of growth and development. They serve a very important purpose; they are the standard against which the students will compare their two sets of experimental plants. The experimental plants will be grown according to the experimental plan that the research team designs. Part of the team's task today will be to begin designing an experimental plan in which one, and only one, variable will be tested.

Figure 2-1

Planning Board for control plants

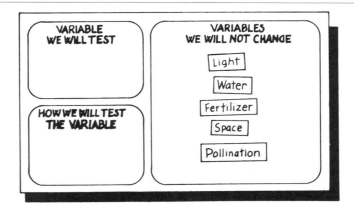

4. Using the Planning Board, demonstrate what each student's Planning Work Sheet (**Activity Sheet 1A**) will look like. First, set up the board in the way it would look for a control plant (see Figure 2-1).

    Then set up the Planning Board to illustrate how it would look for an experiment testing the variable of pollination (see Figure 2-2).

5. Now, have the students break up into their teams. Distribute to students **Activity Sheet 1A, Activity Sheet 1B**, scissors, and a piece of paper. Tell them to work in teams and to follow the directions carefully.

6. Circulate around the room and encourage manipulation and discussion of the variables. Inject such questions as:

    - Which variable would you like to change? Why do you think that would make an interesting experiment?

    - What equipment would you need to carry out that experiment? Could you build it yourself?

Figure 2-2

Planning Board for experimental plants

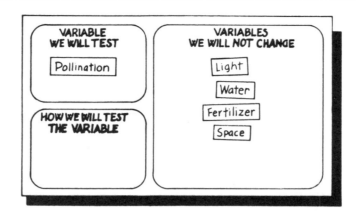

■ What might happen to the plant if you change that variable? What do you think you might find out?

7. Bring the class back together for discussion. Remind the students that they have been working with two very important ideas in science: the fair test and experimental variables. Their task now is to bring the two ideas together in order to come up with a plan for their team's experiment. Emphasize that the experiment should be a fair test and that the way to accomplish this is to change only one variable per experiment and to keep all the rest unchanged. This method ensures the investigation of only one question at a time.

   Note: Although the students are aware that half of the plants will be experimental plants and half will be control plants, they should try not to decide which role each student in each team will play in caring for them. That will be done during the next lesson.

Final Activities

Tell the students that, for the next day or two, they will go into a reflective phase. They need time to think, to let ideas come and go, and to discuss their ideas informally with family, friends, teachers, and each other before deciding on the topic for their experiment. They also need more information. Assign the Reading Selection on pg. 12 of the Student Activity Book. The selection describes the normal life cycle of Wisconsin Fast Plants. It is reproduced on pgs. 25 to 28.

Extensions

1. Encourage the students to do more research on the life cycles of plants and the effects of environmental conditions on plants. Suggest that students report their findings to the class. Topics might include:

   ■ How acid rain is damaging trees

   ■ How salt on roads in northern parts of the country affects grasses

   ■ How pollutants in the air affect lichens growing on trees

   ■ How long it takes plants to recover after a forest fire

2. Encourage the students to locate and read *How to Think Like a Scientist: Answering Questions by the Scientific Method*, by Stephen P. Kramer (see the **Bibliography, Appendix B**).

Reading Selection

Figure 2-3

Growth chart of Wisconsin Fast Plants

The Life Cycle of Wisconsin Fast Plants

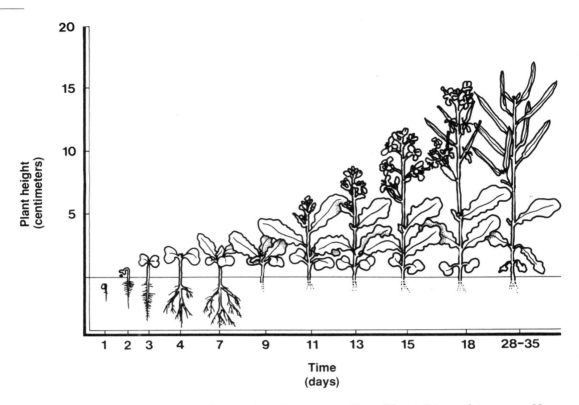

In Lesson 1, you read about what Wisconsin Fast Plants™ need to grow. Now, read about how they grow under ideal conditions. In a remarkably short time, the plants speed through all of the stages in their life cycle.

The tiny root is the first part to emerge from the seed, usually within 24 hours of planting. About 4 hours later, the two **seed leaves** (also known as **cotyledons**) appear above ground. At first, the heart-shaped seed leaves are white, but soon they become green and take on the job of making food for the seedling until the **true leaves** develop fully (at about Day 7).

On Day 4 or 5, the true leaves look like tiny lumps at the growing tip of the plant, between the seed leaves. As they grow, they look quite different from the seed leaves but have the same job: food production. The true leaves continue to emerge over the next several days and grow very rapidly in size.

Soon after the first true leaves have developed fully, the **flower buds** appear (Day 8 or 9). Look for them clustered at the growing tip of the stem. The buds are closed tightly and are greenish-yellow, sometimes tinged with purple. In another day or two, the buds will open to reveal bright yellow, four-petaled *Brassica* flowers.

The second week of life is a time of enormous activity for the plant. Besides growing buds and enlarging leaves, the plant also is going through an upward growth spurt, much as humans do during adolescence. Sometime between Day 9 and Day 13, the plant's stem will grow longer between the places where the leaves are attached.

Identifying Variables and Planning a Fair Test / **25**

Figure 2-4

The growth of leaves and flower buds

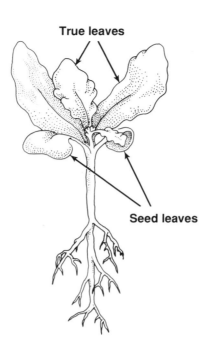
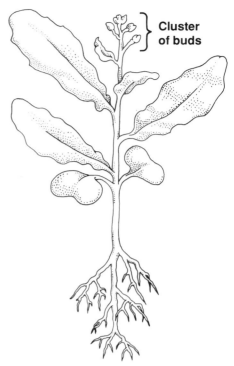

True leaves

Seed leaves

Cluster of buds

From Days 4 through 7, the true leaves continue to emerge and grow at a remarkably fast pace

By Day 8 or 9, the yellow-green flower buds appear in a tight cluster

From Day 13 to Day 18, the flowers are fully open, ready for pollination to take place. In nature, this is best done by a bee. For Wisconsin Fast Plants, **cross-pollination** is necessary in order for the plant to produce fertile seeds. This means that pollen from the male part of one plant must be moved to the female part of another plant. You will read more about pollination in Lessons 6 and 7.

Figure 2-5

The plant's growth spurt

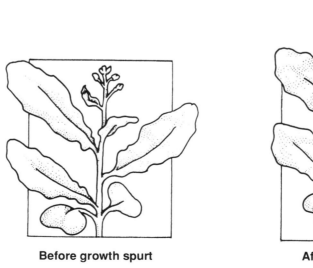

Before growth spurt

After growth spurt

After the transfer of pollen from one plant to another, the flowers begin to change. The petals fade, wither, and fall. The eggs inside the **pistil** are developing into seeds. The pistil, now called the **seed pod**, grows longer and swells as the seeds continue to grow over the next 20 days. During this time, there is very little upward growth because the plant is putting its energy into seed production.

Figure 2-6

A plant with mature pods

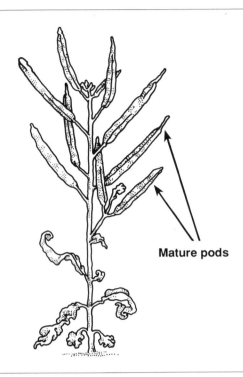

Mature pods

At about Day 35, the plant is removed from the watering system and allowed to dry out so that the seeds may ripen. After about a week, the pods can be snapped off and the seeds removed. The life cycle is complete: the plant has grown from seed, produced its own seed, and now the new seed can be planted to begin a new life cycle.

The *Brassica* plants will grow in this way if they have ideal conditions. Keeping this in mind, what do you think would happen if:

- the plants were shielded from the light for 3 hours a day?
- the plants got no fertilizer?
- the plants got twice as much fertilizer as required?
- the temperature went down to 45°F (7°C) for 6 hours every night?
- the temperature were kept at 90°F (32°C)?
- the plants were overcrowded?
- the blossoms were not pollinated?

Do you think that changing one variable would make a difference in the life cycle of the plant? How might the plants be different? How might their life cycle be slowed down or speeded up? Would they still produce flowers? Would they still produce seeds? What do you think?

Figure 2-7

The life cycle of the
Brassica *plant*

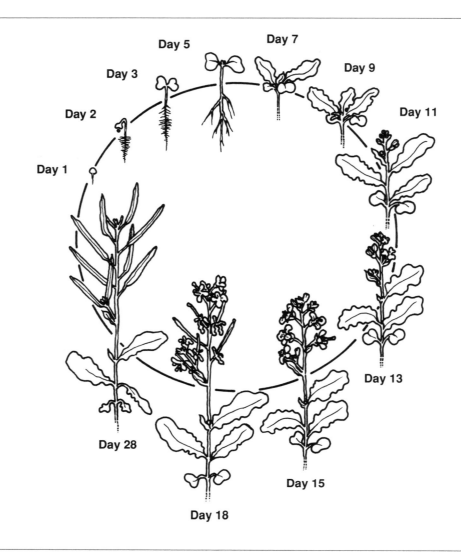

Day 1
Day 2
Day 3
Day 5
Day 7
Day 9
Day 11
Day 13
Day 15
Day 18
Day 28

Planning Work Sheet   **Activity Sheet 1A**

NAME: _____

DATE: _____

What we think will happen:

The variable we will test:

How we will test the variable:

The variables we will not change:

Directions for Using the Planning Work Sheet   **Activity Sheet 1B**

NAME: _____

DATE: _____

Directions: Do each of the following activities as a cooperative research team. Be sure that you have **Activity Sheet 1A, Planning Work Sheet** before you begin working.

☐ 1.   Cut out pieces of paper that measure approximately 1" x 3". Using a separate piece of paper for each variable, write out each of the requirements that Wisconsin Fast Plants need to grow well.

☐ 2.   Place the pieces of paper with the variables written on them one under the other in the right-hand column of **Activity Sheet 1A**, under the heading, "Variables we will not change." These variables represent the ideal growing conditions for Wisconsin Fast Plants. The plants that are "normals" or "controls" will grow under these ideal conditions.

☐ 3.   As a team, select one variable that you plan to change during your team's experiment. Discuss it with your teammates. Then put that piece of paper in the left-hand column of **Activity Sheet 1A**, under the heading, "Variable we will test." (None of the other variables will change; leave them where they are.) What are the ways this variable might be changed to make a good experiment? Talk about this, and share your ideas about these questions:

  ■ Would changing this variable make an interesting experiment? Why?

  ■ Would changing this variable make a doable project? Would we have any special problems with materials or equipment? Could we handle these problems?

  ■ What might happen to the plant if we change this variable?

☐ 4.   Place the piece of paper with the variable you have just considered back in the right-hand column. Now, select another variable to discuss. Put that variable under the heading, "Variable we will test." Repeat the process in Step 3 above.

☐ 5.   Continue reviewing the variables in this way until your teacher calls you together for a class discussion.

☐ 6.   Save this sheet for the next lesson.

Outlining the Experimental Plan

Overview

In Lesson 2, the students began discussing how to design a controlled experiment, one that would be a fair test. By using the planning board, the students discovered that only one variable may be changed at a time in an experiment; all of the others must be kept constant in order for the experiment to be a fair test.

In this lesson, the students make specific choices about their experimental plans. They decide which variable they will alter and how they will alter it. In addition, students decide who will grow control plants and who will grow experimental plants. Finally, they also are introduced to the importance of keeping records throughout a scientific experiment.

Objectives

- The teams decide on a specific topic for their experiments.

- Each team designs an experimental plan.

- Each student begins a recordkeeping journal.

Background

Summary of Experiments

Four types of experiments work well using Wisconsin Fast Plants™. These are experiments dealing with the variables of fertilizer, space, light, and pollination. Experiments dealing with water or temperature are very difficult to control and, therefore, are not recommended for elementary school students.

Each of the four types of experiments is described below with scheduling notes, examples of topics that students have developed successfully, and procedures. A chart summarizing these four experiments and keyed to the lessons during which each experiment would start is given on pg. 33.

Ideally, it will be best for your students to devise their own questions and their own experimental plans. The experiments described here are meant to be only illustrations of topics that have worked.

Variable to be tested: FERTILIZER

Scheduling: The experiment begins on planting day (Lesson 4), which is when the fertilizer is placed in the quads.

Successful topics: 1) Will the experimental plants grow taller (have more leaves, produce more seeds) if we give them twice the recommended amount (or three times, or four times, etc.) of fertilizer? or 2) Will the experimental plants be shorter than the control plants if we do not give them any fertilizer at all?

Procedure: On planting day, the students place the amount of fertilizer called for by their experiment into each cell of the experimental quads. The control quads are planted with the recommended three pellets of fertilizer per cell.

Control of variables: All of the quads are exposed to the lights for 24 hours a day, receive the same water, and are thinned to one plant per cell. They share the same temperature and all of them are cross-pollinated.

Note: Many commercial potting soil mixes contain a certain amount of fertilizer themselves. This type of mix can significantly reduce the effect of omitting the fertilizer pellets.

Variable to be tested: SPACE

Scheduling: On planting day (Lesson 4), students begin to implement their experimental plan either by overplanting or by planting the standard two seeds, but then they do not thin out the plants on Day 4 or 5 (Lesson 5).

Successful topic: If we do not thin out the experimental plants to one plant per cell, but leave them overcrowded with two (or more) plants per cell, will they be shorter (or have fewer leaves or produce fewer seeds) than the control plants?

Procedure: Two approaches are: 1) Overplanting on planting day. Three seeds are put into each experimental cell instead of the standard two. The seedlings are not thinned out or transplanted. 2) Planting the standard two seeds per cell, but deliberately leaving the experimental quads overcrowded at two plants per cell. Do not thin out or transplant any of the seedlings.

Control of variables: All of the quads are exposed to the lights for 24 hours a day, receive the same water, and are given three fertilizer pellets per cell at planting time. The temperature is the same for all, and they are all cross-pollinated.

Variable to be tested: LIGHT

Scheduling: Begin excluding light from the experimental plants a day or two after planting (between Lessons 4 and 5).

Successful topic: What will happen to the number of leaves (or the height of the plant) if the experimental plants are in the dark for 5 hours (or 4 hours, or 3 hours) a day every school day?

Procedure: The students construct lightproof boxes from milk cartons (tinfoil tents would work also) and place them over the experimental quads of plants without removing them from the watering system.

Control of variables: The students place the control quads of plants under the lighting system next to the darkened experimental plants so that all of the sets will share the same water and the same temperature. All of the plants get three fertilizer pellets each. All are cross-pollinated. All are thinned to one plant per cell.

Note: Covering plants under light may also cause the temperature to rise under the covering

Variable to be tested: POLLINATION

Scheduling: Cross-pollination needs to be done for 5 consecutive days after the flowers open, or from about Day 12 to Day 18 (see Lesson 7).

Successful topic: Will a plant produce seeds if it is not pollinated?

Procedure: The experimental quads of plants are not pollinated. The students also take measures to prevent accidental pollination. They construct barriers (stiff paper or index cards work well) and place them between the four experimental plants in each quad. They also could place each quad inside a milk carton with the top cut off for light, so that each experimental quad is isolated from the others as much as possible.

Control of variables: All plants are under the lights 24 hours a day and are living in the same temperature, receive the same water, are given three fertilizer pellets each at planting time, and are thinned to one plant per cell.

Summary of Four Major Experiments

Variable to Be Tested	Variables Not Changed	Procedure	Scheduling
FERTILIZER	Space Light Pollination Water Temperature	Follow all the standard directions except: Change the amount of fertilizer per cell to either more or less than 3 pellets	Experiment begins on planting day (Lesson 4).
SPACE	Fertilizer Light Pollination Water Temperature	Follow all the standard directions except: Plant 2 seeds but do not thin OR Plant more than 2 seeds and do not thin	Experiment begins on planting day (Lesson 4). Do not thin or transplant (Lesson 5).
LIGHT	Fertilizer Space Pollination Water Temperature	Follow all the standard directions except: Deprive the plants of light for part of 24 hours each school day	Construct light blockers before Lesson 4. Put them in place after planting. Manipulate each school day. Consider how to handle weekends

POLLINATION	Fertilizer Space Light Water Temperature	Follow all the standard directions except: Do not make a bee stick and do not pollinate. Instead construct pollination-prevention devices	Construct and insert devices about Day 10, 11, or 12 before flowers open (Lesson 7). Leave devices in place until harvest (Lesson 8).

Materials

For each student

1   student notebook
1   **Activity Sheet 1A, Planning Work Sheet** (from Lesson 2)
1   **Activity Sheet 1B, Directions for Using the Planning Work Sheet** (from Lesson 2)
1   **Daily Data Record** (optional)

For each four-member team

1   **Activity Sheet 2, Outlining the Team's Experiment**
    Glue

For the class

1   **Planning Board** (from Lesson 2)

Preparation

Duplicate the **Daily Data Record** and **Activity Sheet 2**.

Procedure

1. Ask the teams to take out **Activity Sheets 1A** and **1B** from the previous lesson. Tell them that, today, each team will decide which variable to test in its experiment. The Student Activity Book explains in more detail how to fill out these sheets.

2. Each team must make another important decision today about how to manage its four quads of plants once the experiment has begun. There are several alternatives to discuss with the class:

    ■   Each person on the team could be exclusively responsible for one set of four plants (one quad). This means that the team would have to divide itself into two students solely responsible for the two sets of experimental plants and two students solely responsible for the two sets of control plants. It would be understood that teammates would share information with each other and fill in for each other in case of absence.

        The advantage of this arrangement is that each student takes immediate ownership of one set of plants. The disadvantage is that students designated as caretakers and recordkeepers for the control plants sometimes feel that they are not participating fully in the experimentation. Both sets of students need to be made sensitive to the advantages and disadvantages of this arrangement and to the need to work on the project cooperatively.

    ■   After dividing up the caretaking and recordkeeping responsibilities as outlined above, team members could plan to switch roles halfway through the experiment. That is, those who were designated as caretakers and recordkeepers for the control plants could change places with their teammates who were responsible for the experimental plants.

■ The team as a whole could share responsibility for the care and recordkeeping of all four sets of plants. The advantage of this arrangement is that the students can work out a rotating system in which each person works with both the control plants and the experimental plants throughout the length of the experiment. But this plan also requires constant renegotiation of duties and greatly complicates the process of recordkeeping.

3. The next challenge that each team faces is deciding which question or problem to investigate. Discuss this problem as a class. It is important that the students develop a question with a narrow focus so that they investigate a specific topic. For example, a good question is, "What happens to the number of seeds a plant produces if we double the normal amount of fertilizer?" A question that is too broad or too vague, such as, "What happens if we give the plant a lot of fertilizer?" will be difficult to answer.

Use the illustration on pg. 18 of the Student Activity Book (Figure 3-1 in the Teacher's Guide) to help the students formulate good, specific questions.

Figure 3-1

How could each of the variables affect the plants?

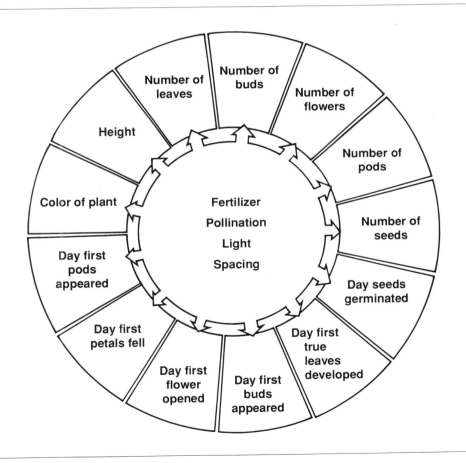

4. Distribute to each team **Activity Sheet 2: Outlining the Team's Experiment**. Allow the teams time to discuss the specific questions they

might investigate. Once they have decided on a question, they should record it on the activity sheet.

5. Now the teams should complete the rest of **Activity Sheet 2**. It will help them to define their experimental plan more clearly. Before the end of the class, collect each team's completed outline sheet for review.

6. Before the end of the class, distribute a copy of the **Daily Data Record** on pg. 37 to each member of the class and tell them to clip it into their notebooks for later use. Or, use the sheet as a model to show the students how to set up their notebooks.

7. To set the stage for the daily recordkeeping, discuss briefly the importance of recordkeeping and preview the kinds of information that might be included on the **Daily Data Record**. For example, daily data could include a sketch of the plant on that date along with a written description of its appearance (height, color, number of leaves and buds), a note of any changes or new developments, or a "no-change" notation.

    Things that can be measured or counted are also of great importance. These include the number of seeds that germinate, the number of leaves, buds and flowers, and the length of leaves. The time it takes for developments to occur in the life cycle should be noted also. The dates of germination, appearance of true leaves, first flowering, and pod formation are all significant.

8. Emphasize that every student is expected to keep records every school day. Try to allow time at the end of class for this activity. Stress the importance of dating all entries.

Final Activities

The next class is planting day. Tell the students to preview the directions on planting found on pg. 19 of the Student Activity Book. These are the same directions they will be following from **Activity Sheet 3** of Lesson 4, when they actually plant.

Evaluation

Before planting day, take time to review each team's experimental plan as given in **Activity Sheet 2: Outlining the Team's Experiment**. Unless the students understand the elements of a good experiment, the rest of their experience with this unit will be disappointing.

While reviewing each team's **Activity Sheet 2**, make sure that the students have done the following:

■   Chosen one variable to change and listed the other variables as constants

■   Identified a specific testable question involving that variable

■   Identified something to measure

■   Identified any foreseeable difficulties with equipment

■   Chosen a project that is possible for them to do.

■   Identified some observable features to monitor, such as height, color, time of flowering, and number of leaves and pods.

Note: If necessary, delay planting until you have given guidance to the teams that need it.

Daily Data Record

My plants are:
☐ control plants
☐ experimental plants

NAME: _____

DATE: _____

Plant 1	Observations	Measurements	Sketch

Plant 2	Observations	Measurements	Sketch

Plant 3	Observations	Measurements	Sketch

Plant 4	Observations	Measurements	Sketch

Outlining the Team's Experiment   **Activity Sheet 2**

NAME: _____

DATE: _____

1. The one variable we will test is: _____

2. The question we will try to answer is:

3. In order to make our experiment a fair test, we will keep all of these variables constant (unchanged):

    1) _____

    2) _____

    3) _____

    4) _____

    5) _____

4. Will we need some special materials or equipment? _____ If so, what will we need?

    _____

    _____

5. What we will measure: _____

    _____

    _____

6. What we will count: _____

    _____

    _____

7. What we will observe: _____

    _____

    _____

8. What we think will happen to the experimental plants: _____

    _____

    _____

Planting the Seeds

Overview

During this lesson, each student plants one set of seeds in a four-celled container called a quad. How the quad is planted depends on the experimental plan that the student's team has devised. The success of the rest of the project depends on everyone planting correctly today.

Teams experimenting with fertilizer and teams experimenting with overcrowding will begin manipulating their variable today; therefore, they will plant according to their own set of instructions. Teams experimenting with variables other than fertilizer and overcrowding will plant according to the standard instructions on pg. 46.

Because of the brassica plant's unique characteristics, certain planting methods and materials are required. They are explained in detail in this lesson. For the best results, planting day should be a Monday or a Tuesday.

On the day before planting, you may want to take the class through Steps 1 and 2 of the **Procedure** in this lesson as an introductory session.

Objectives

- Students learn to plant.
- Students experimenting with fertilizer and with overcrowding begin manipulating their variables.

Background

The success of this lesson depends on organization and preparation. Four important setups must be completed before the students can begin planting. You may wish to recruit another adult or two to help.

The setups are:

- constructing the lighting system (explained in **Teaching Strategies and Classroom Management Tips**, with additional instructions in the *Experiments with Plants* materials kit)
- arranging the watering system (explained in the **Preparation** section on pg. 41)
- organizing the planting materials for distribution (explained in the **Preparation** section)
- preparing the work spaces and the cleanup area (explained in the **Preparation** section)

Materials

Be sure to plant extra quads of plants. You might need them if students drop their plants, if they are absent on planting day, or if new students join your class.

For each student

    1 student notebook

For every two students

    1 pair of forceps
    1 cup of water
    1 dropper

For each student planting according to the standard instructions

    1 **Activity Sheet 3, How to Plant Wisconsin Fast Plants™ Seeds: Instructions and Checklist**
    1 tray for carrying supplies
    1 planter quad
    1 5-ounce cup of potting mix
   12 fertilizer pellets
    1 toothpick
    1 paper towel
    1 spoon
    4 wicks
    8 seeds (2 per quad cell)
    1 planter label

For each student experimenting with fertilizer

These students will need all the standard material and should follow the standard instructions **EXCEPT** that they should use only the amount of fertilizer called for by their team's experimental plan. The amount will be either more or less than the standard amount, which is twelve pellets.

For each student experimenting with overcrowding

These students will need all the standard material and should follow the standard instructions **EXCEPT** that they should use only the number of seeds called for by their team's experimental plan. This amount could be eight seeds (the standard amount), in which case the team will not thin out the plants in Lesson 5, and will leave them overcrowded. Or the amount to be planted could be more than eight seeds in a quad.

For the class
Cleanup supplies:
6 to 8 sponges
    2 dishpans of water
    1 plastic-lined trash can

1   dustpan and whisk broom
1   overhead projector and screen (optional; for demonstration of the
    differences between the fertilizer pellet and the plant seed)

Preparation

1. Duplicate **Activity Sheet 3**, one per student.

2. Obtain the cleaning supplies not included in the materials kit.

3. Explanations for arranging the watering system, organizing the
    distribution station, preparing work spaces, and setting up the cleanup
    area are given below. Please see **Teaching Strategies and Classroom
    Management Tips** and the instructions in the *Experiments with Plants*
    materials kit for guidance on how to set up the light bank.

4. If necessary, obtain and set up the overhead projector and screen.

The Watering System

The Wisconsin Fast Plants' unique watering system delivers water from the
tank by capillary action. The water travels from the tank through the mat,
then through wicks to the potting mix. Below are instructions for setting up
the watering system. Figure 4-1 shows how the watering system works. Once
the system is set up, all you need to do is to refill the tank every 4 to 5 days.

Figure 4-1

*How the watering
system works*

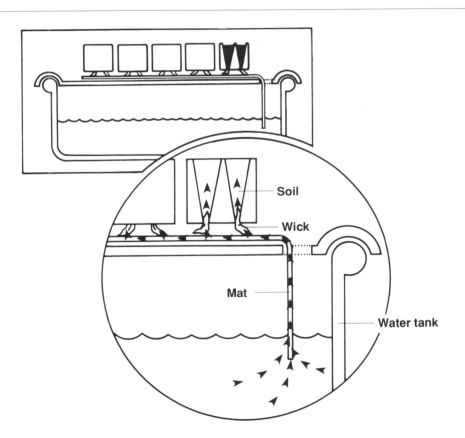

- Fill the water tank to capacity.

- Place the water mat in the tank to soak for about 15 minutes. Squeeze the mat out once or twice during this time and place it in the tank again so that it is saturated.

- Without squeezing the mat out, lay it on top of the tank's lid with the end dangling in the water. Smooth out any air pockets.

- Drop the blue anti-algal (copper sulfate) squares into the tank to prevent algae from growing and clogging the system.

- After planting, make sure that each quad rests squarely on the water mat so that the wicks at the bottom of each quad cell come into contact with the mat. Check the water mat and the soil in the quads each day. Both should be wet. This indicates that the wicking system is working. Water from the top if the soil appears dry.

The Distribution Station and the Materials

The distribution station needs to be set up with all of the materials for planting. Arrange the materials "cafeteria style" so that the students can pick up each item they need. This method saves time for teachers and is a learning experience for students. To set up the distribution station efficiently, follow the illustration in Figure 4-2 and the guidelines on pg. 43.

Figure 4-2

The distribution station

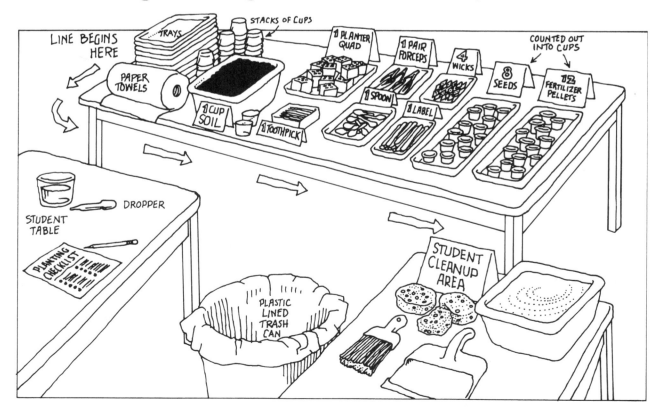

■ Select one large area or several small areas of the room where the students can walk by easily in single file on both sides of the supplies.

■ Position all of the materials in a line on a series of desks, tables pushed together, or on the floor, if necessary.

■ Ward off bottlenecks in the line by counting out the twelve fertilizer pellets and the eight seeds ahead of time and placing them in small cups.

Note: These are the standard numbers of pellets and seeds for those teams NOT experimenting with fertilizer or overcrowding. Those who are experimenting with the fertilizer and overcrowding variables need to adjust the numbers according to their experimental plans.

■ Moisten the potting mix if it has become powdery.

■ Empty the potting mix into a large container. Have the students dip their cups into this container.

■ Place a label on each item telling students what it is and how many to take.

The Work Spaces

To help the students work together more efficiently, organize them as follows:

■ Allow the teams of four to work together so that they can share watering supplies, forceps, and ideas easily.

■ Clear the work area of anything that might get wet or dirty.

■ Have four copies of **Activity Sheet 3, How to Plant Wisconsin Fast Plants™ Seeds: Instructions and Checklist** for each team at their work space, one per team member.

The Cleanup Area

Place the cleanup supplies—basins of water, sponges, whisk broom, dustpan, and plastic-lined trash can—in a part of the room where they will be noticeable, but out of the way.

It may be wise to discuss cleanup procedures. Let the students know that they will be expected to clean up thoroughly, independently, and cheerfully.

Procedure

1. Distribute **Activity Sheet 3, How to Plant Wisconsin Fast Plants Seeds: Instructions and Checklist**. Preview it with the class. Instruct all the students who are planting experimental quads to fill in one of the two blank boxes on the instruction sheet with the correct numbers of fertilizer pellets and seeds for their experimental plan. Remind them that the standard numbers are twelve fertilizer pellets (three per cell) and eight seeds (two per cell, even if you are not overcrowding, since all the seeds may not germinate).

2. Ask the group experimenting with the fertilizer variable to give the class a very brief description of its project. The group's members should have identified what will constitute the control (the standard amount of fertilizer, three pellets per quad cell) and what amount of fertilizer they will add to each experimental quad section. The rest of the class can

make predictions about what they think will result from over- and underfertilizing.

Ask the group experimenting with the overcrowding variable to give the class a very brief description of their project. They should have identified what will constitute the control (the standard number of seeds, two per quad cell, later thinned to one plant per section) and how many seeds they will place in each experimental quad cell. The rest of the class can make predictions about what they think will result from overcrowding the plants.

3.   Now, focus attention on the distribution station. Demonstrate how to walk carefully through the line, read the labels, wait for your turn, and pick up the correct number of each item. Have the students pick up their supplies following your model.

     Note: Students will have to share some items (such as water and droppers), but each student will be planting one four-celled quad independently, following the instructions in **Activity Sheet 3**.

4.   It is time to plant. Tell the students to use the illustrations on **Activity Sheet 3** on pg. 46 to guide them (see Figure 4-3, also). Circulate so that you can help as needed. This is an opportunity to assess informally how well the students are following directions.

5.   As work progresses, check these trouble spots:

     ■   Be sure that the fertilizer goes in first, not the seed. To point out the difference in size, lay both of them on an overhead projector, if available.

     ■   Seeds planted too deeply will not germinate, or sprout, properly.

     ■   A great splash of water will wash out the seed. Each section must be saturated gently and thoroughly in order for the wicks to start carrying up water.

     ■   Check that all plants are sitting squarely on the water mat before leaving for the day.

6.   Cleaning up today is extensive and requires everyone's help. The more specific your instructions, the better your students will perform. Decide ahead of time exactly how you want the room to look at the end of the activity, and let the class know that nothing less will do. Encourage everyone to do a fair share of the work. Make sure that students return their supplies to the distribution station.

7.   Allow time for the students to make journal entries. Suggest that they include the date, a brief description of what they did, and any problems they had.

Final Activities

Alert the group experimenting with hours of light that their experiment should be ready to start in the next day or two. Have the members describe their project very briefly to the class. They should have planned how many hours of light the experimental plants will get each day, how the experimental plants will be shielded from light, what to do with the control plants, and what they will do about weekends.

Ask the class to make predictions about the effects that less than 24 hours of light will have on the experimental plants. Encourage students to offer suggestions to this group.

Note: For the next 3 days (or until you are positive that the wick system is working), the planters must be watered from the top. If the potting mix in the planters appears dark and feels moist when you arrive in the morning, you can be sure that the wicks are carrying water up into the plants. After that, all you need to do is refill the water tank every 4 or 5 days.

Figure 4-3

The planting sequence. Numbers correspond to steps on Activity Sheet 3.

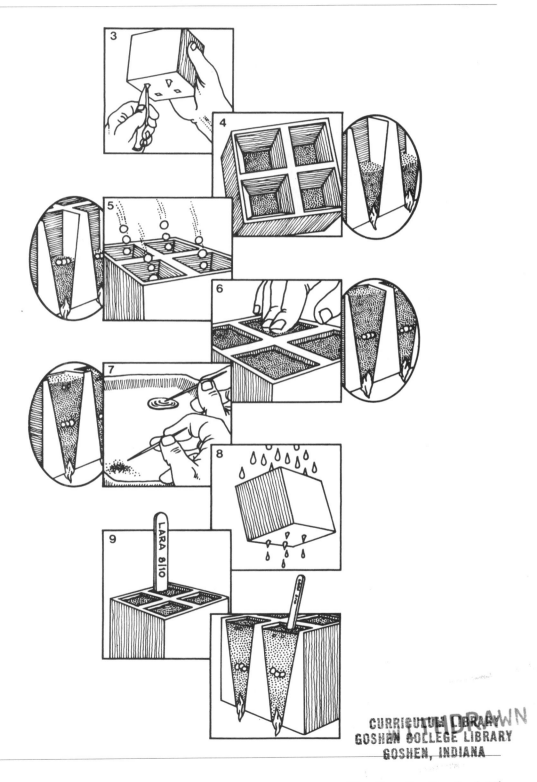

How to Plant Wisconsin Fast Plants™ Seeds: Activity Sheet 3
Instructions and Checklist

My plants are: NAME: _____

    ☐ control plants   DATE: _____
    ☐ experimental plants

The brassica plants are special in many ways so you must follow certain directions when planting the seeds. It is very important to follow the directions carefully. Do one step at a time. Check off each step when you finish it.

☐   1.   Pick up all of your **supplies** from the distribution station. Be sure that you and your partner have these items before you begin planting:

____	1	tray	____	8	seeds*
____	1	planter quad	____	1	toothpick
____	1	spoon	____	1	label
____	1	cup of soil	____	1	paper towel
____	4	wicks	____	1	pair of forceps
____	12	fertilizer pellets*			

        *If you are experimenting with **fertilizing** or **overcrowding**

         ____☐   fertilizer pellets
         ____☐   seeds

☐   2.   Number each **cell** of the planter quad from 1 to 4.

☐   3.   Place one wick in each cell of the planter quad. Use your forceps to pull the wick through the hole until the tip sticks out about 1 centimeter (¼").

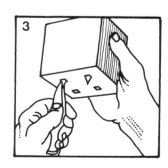

☐ 4.   Using the spoon, fill each section of the planter quad halfway with **potting mix**.

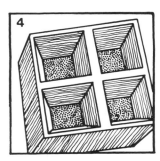

☐ 5.   If you are not experimenting with fertilizer, add three **fertilizer pellets** to each cell. Look closely. The fertilizer pellets are much larger than the seeds.

*If you are experimenting with fertilizer, add to each cell the number of pellets called for in your experimental plan.

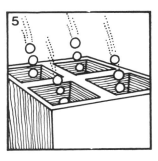

☐ 6.   Fill each cell to the top with **potting mix**. Press it down a little with your fingers.

☐ 7.　Put a drop of water on your tray and dip your toothpick in it. Use the wet toothpick to pick up one **seed**. Place the seed just below the top of the potting mix in one cell and cover it. Plant a second seed in this cell in the same way. Repeat until there are two seeds in each cell of the planter.

*If you are experimenting with overcrowding, plant in each cell the number of seeds called for in your experimental plan.

☐ 8.　Using the dropper, **water** very gently, a drop or two at a time, until water drips from the bottom of each wick.

☐ 9.　Write your name, today's date, and either E (for experimental) or C (for control) on the **label** and place it in the planter.

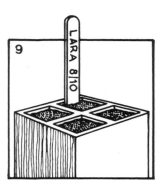

☐ 10. Place your quad under the lighting system with the label facing out. Double-check to ensure that your planter is completely on the **water mat**.

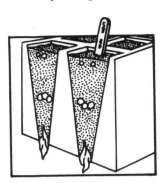

Thinning and Transplanting (Day 4 or 5)

Overview

Today, most students will thin their plants to one per cell, the ideal condition. Only teams experimenting with overcrowding will not thin out their plants. Their quads will remain overcrowded, according to their team's experimental plan.

Objectives

- Students learn why thinning and transplanting are sometimes necessary for the health of a plant.

- Students gain practical experience with these two gardening techniques.

- Students experimenting with overcrowding continue their manipulation of their variable by allowing their plants to remain overcrowded.

Background

Your students will balk at the idea of thinning; even the most seasoned gardener dreads it. The excitement of planting and of watching the seedlings emerge is still so fresh. Who wants to pull up those tender young sprouts?

But thinning is important. As gardeners thin their beds to give their plants the best possible growing conditions, so must we. For the *Brassica* plants, it is best to thin to one plant per cell. This ensures that each plant will have ample space, light, food, water, and air circulation. Under these conditions, the plants will thrive and will yield the highest possible number of seeds at harvesttime. Plants grown under more crowded conditions must compete for these essentials and will be less productive.

There are several ways to thin. One way is to pinch off the plants at the soil line or to cut them off at the soil line with scissors. The advantage to this method is that it will not damage the roots of the plant left behind. Another way is to pull up the plant, with its root system attached, and transplant it to another location, perhaps into a cell where no seeds germinated.

Transplanting requires planning ahead and a gentle touch. Be sure that each student has decided where the plant is to go before he or she uproots it. Minimizing the plant's time in transit is important in order to prevent root damage. The *Brassica* is a survivor, however. Most transplants probably will perk up a few hours after being placed in their new cells, although those plants may lag behind developmentally by a few days.

Materials

For each student
  1 student notebook
  1 toothpick
  1 sheet of graph paper (optional) (A graph paper black line master is
     provided at the end of the lesson on pg. 55.)

For every two students
  1 hand lens
  1 pair of forceps

Preparation

1. Duplicate the graph paper on pg. 55, if needed.

2. If you need extra plants, prepare at least four quads in advance to
    receive the leftover seedlings. Follow all of the steps on **Activity Sheet 3**
    from Lesson 4, except for adding seeds.

Procedure

1. Open the discussion by asking if anyone in the class has had experience
    thinning or transplanting. Follow up with a question asking why it is
    important to thin plants. For example, why does a gardener thin out a
    row of seedling carrots? Then discuss why sometimes it is necessary to
    transplant. Ask: "Why would you transplant a sun-loving rose bush to a
    new location?" Help the students see that the purpose of both
    techniques is to improve growing conditions for the plants.

2. Make sure that everyone knows what they are supposed to do today.
    Then, identify which team (or teams) is experimenting with the variable
    of overcrowding. Ask them to explain once again, very briefly, why the
    experimental plants will not be thinned. Also, ask them to explain how
    many seedlings they will be keeping in each cell and how the control
    plants will be grown. Ask, too, what they expect the outcome of
    overcrowding will be.

3. Ask the students to retrieve their plants and to spend a few minutes
    observing them with a hand lens. Point out differences among the plants
    even at this very early stage of development. Ask, "Are all of your
    seedlings the same size? The same color? Where are the differences,
    exactly? In the shape or size of the leaf? In the length of the stem? Did
    every seed sprout, or germinate?"

4. Students NOT experimenting with the overcrowding variable should
    follow the steps listed and illustrated below to thin and transplant the
    plants.

    ■ Students should select one plant per cell (a total of four plants per
       student) that they will keep; they should thin out the rest.

    ■ Before thinning, students should loosen the potting mix gently with a
       toothpick. Have them poke a hole about 2 cm deep (the pencil point
       up to the paint) in the soil of the container where the seedling will go.
       They can use forceps or fingers to lift out the seedling by its roots,
       and place it in the hole they have prepared. Caution students to
       work gently to avoid crushing the delicate stem. Then they should
       pat the potting mix down a little around the roots of the newly
       transplanted seedlings.

Figure 5-1

How to thin plants and transplant them

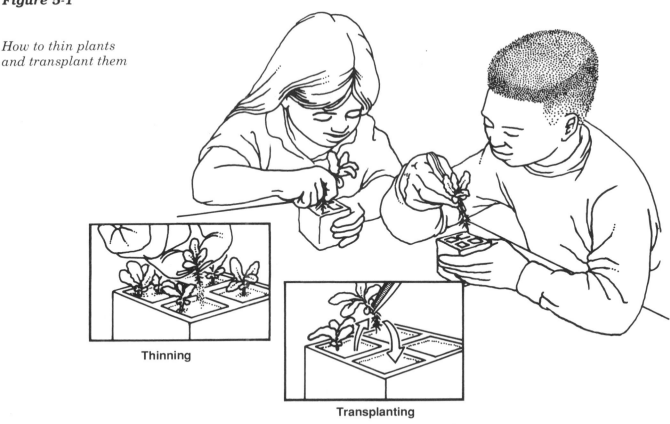

Thinning

Transplanting

5. Students have the following choices of what to do with the uprooted seedlings:

- Transplant them into one of their own cells where seeds have not germinated.

- Donate them to a classmate or place them in the extra quads you prepared (see **Preparation**, No. 2).

- Set aside one seedling to observe. See the **Extensions** at the end of this lesson for more details.

6. Clean up. Tell the students to discard any plants that cannot be used and to return equipment to the proper containers.

Final Activities

With this lesson, students begin to record their observations in their notebooks. Take a moment to discuss the best ways to document today's activities and observations. Ask the students to consider all the possibilities—drawing, writing, measuring, counting, and graphing—and to use as many of them as needed to record a complete and accurate picture of their plants in their individual notebooks. Drawings will prove to be particularly useful in assessment. See the **Evaluation** section at the end of this lesson.

This is a good time to distribute the graph paper. Students may use it to begin recording the height of their plants.

Extensions

Encourage the students to use their hand lenses to make close observations of the root system of one of their uprooted seedlings. This is the only time that the roots will be exposed.

Figure 5-2

Tap root of the Brassica *plant and root hairs*

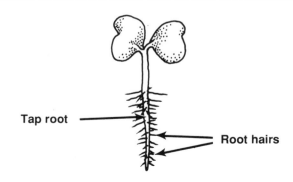

Young *Brassica* seedlings taste surprisingly good, much like broccoli. You might allow students to wash and eat some of the extra seedlings.

Evaluation

In this unit, students are learning how to observe. Their drawings are a good source of information about what they are seeing. It will be useful to compare these early drawings to later ones.

Graph Paper

NAME: _____

DATE: _____

Getting a Handle on Your Bee

Overview

For the next two lessons, students will be focusing on **pollination**, the beginning of the process through which plants are fertilized. They also will be learning about the complex relationship between the worker honeybee and flowering plants. In this lesson, students learn about the physical characteristics of the honeybee. They observe real bees that have died naturally and have been dried. They attach them to sticks so that they are ready for use in pollination in the next lesson. Students also use magnifiers to observe any blossoms that might have opened.

The team members experimenting with the pollination variable will not need to make bee sticks because they will not be pollinating their plants. Instead, after making observations of the dried bees, they should read the information on pollination entitled *The Bee and the* Brassica: *Interdependence* in Lesson 7, pg. 40 of the Student Activity Book (pg. 70 of the Teacher's Guide). Then these students should devise and construct barriers to prevent accidental pollination.

Objectives

- Students learn about the anatomy of the honeybee and the *Brassica* flower through close observation with a magnifier and through reading.

- Students who will be pollinating follow instructions to make bee sticks.

- Students who will not be pollinating gain exposure to pollination through reading; they also devise and construct pollination prevention barriers.

- Students continue keeping daily records of their plants.

Background

Despite the many benefits derived from the work of honeybees, many of us are afraid of them or, at best, think of them as pests to be warded off before they sting. Bees do not deserve such bad press. They are vital contributors to life on earth. In addition to producing wax and honey, the bee is a major agent of pollination, the process by which **pollen** is transferred from the male part of one plant to the female part of another plant. This allows **fertilization** and seed production to take place. (For more information about pollination, see the **Background** in Lesson 7, pg. 65 and the Reading Selection on pg 70.)

Undoubtedly, you will discover the full range of attitudes toward bees in your class. At first, your students will express all kinds of negative reactions, and you may share some of their sentiments. This is normal.

As your students learn more about bees, a transformation will take place. The students will find it exciting to observe this otherworldly creature close up, especially now that it is harmless. The noise level will go up, but, if you listen carefully, the conversation probably will be "bee"-related.

After pollination has been introduced and the students begin to use their bee sticks as tools, they will handle the creatures quite matter-of-factly. Many will begin to take pride in how well their bee stick works. By the end of the unit, some students will ask to keep their bees!

Students may be curious about where the bees they are working with came from. They are real and died a natural death outside the hive at summer's end.

The Colony

The honeybee is a social insect that lives in a remarkably well-organized colony consisting of three kinds of bees: the **queen**, the **drones**, and the **workers** (the kind used in this unit). Each kind of bee has basically the same three-part body plan consisting of the **head**, the **thorax**, and the **abdomén**. Because each kind of bee has a different job to do, parts of their bodies have evolved in specialized ways. Below are illustrations and brief descriptions of the bees in the colony and the jobs they do. These illustrations are suitable for reproduction as an overhead transparency. The illustration on pg. 63 shows a worker bee with its body parts labeled.

Figure 6-1

Three kinds of bees in a colony

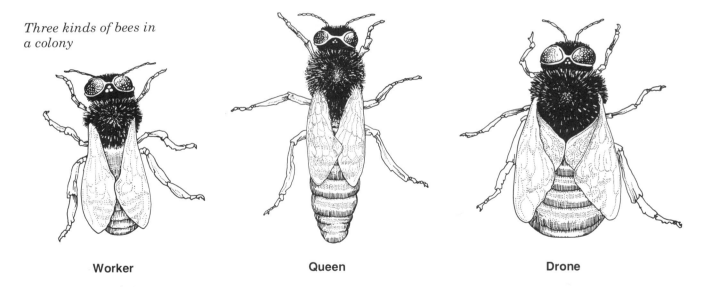

Worker   Queen   Drone

QUEEN   Life span: 3 to 5 years

The largest occupant of the hive and the only one of her kind, the queen bee is a virtual egg factory, capable of producing about 1,500 eggs a day. Shortly after hatching, the virgin queen takes one nuptial (mating) flight and is fertilized for life by a drone, with the drone's sperm being stored in a special sac in her body. Then the queen returns to the hive to begin laying eggs that will become either workers or drones.

DRONE   Life span: 1 or 2 seasons (spring or summer or both)

Stockier than the queen and a strong flier, the drone makes up about 10 percent of the population of the hive. His only purpose in life is to catch the queen during the nuptial flight and fertilize her. Ironically, the winner dies in the act. The rest of the drones return to the hive to be fed and cared for by the worker bees until food gets scarce in the fall. Then the workers bite off the drones' wings and unceremoniously throw them out into the cold to die.

WORKER   Life span: 3 to 6 weeks

Smallest in size but comprising 90 percent of the hive's population, this bee is always a sterile female. The hive could not exist without her and she literally works herself to death. At different stages in her life, she specializes in different tasks such as feeding larvae; building, cleaning, and guarding the hive; secreting wax; controlling the hive's temperature; and collecting nectar and pollen.

The body of the worker bee is specially adapted for collecting food. Her long strawlike tongue siphons up **nectar** from deep inside the flowers. The nectar is stored in a nondigesting **honey stomach** for transport back to the hive. The worker's hairy body traps pollen that the bee stores in a pollen basket on its hind leg. In this manner, both the nectar and the pollen are carried back to the hive to feed the colony.

Materials

For each student pollinating

1   student notebook
1   dried bee
1   toothpick
1   tray
1   hand lens
1   **Activity Sheet 4, How to Make a Bee Stick**

For each four-member team pollinating

1   small cup of white glue
1   paper cup

For each four-member team not pollinating

1   dried bee
1   hand lens
    A variety of recycled materials from which to construct pollination prevention barriers for their experimental plants. Suggestions include large index cards, oak tag or other stiff paper, corrugated cardboard, milk cartons, and tinfoil.
    Reading materials on pollination

For the class

1   overhead transparency of "The Worker Bee's Body"
    (see **Appendix C**, pg. 138)
    Overhead projector and screen

Preparation

1. Duplicate **Activity Sheet 4**.

2. Place all the supplies (except the **Activity Sheet**) in the distribution station for each student to pick up. Remember that four to six students can be grouped to share cups.

3. Put a small dollop of glue in a cup at each work table. Place an inverted cup at each work table into which the students can poke their bee sticks after they have finished making them.

4. Obtain and set up the overhead projector and screen.

Procedure

1. Distribute **Activity Sheet 4** and preview it quickly with the class. Explain that most students will be pollinating their plants during the next class (and for 5 days thereafter), so they will need to make bee sticks. Students who are experimenting with the pollination variable by not pollinating will not need to make bee sticks, but they will be expected to observe the bee.

2. Students making bee sticks should follow the directions listed below:

   ■ Have students pick up their supplies and make a bee stick according to the directions on **Activity Sheet 4**. Encourage them to work independently.

   ■ Tell students to use their hand lenses to examine the bee's body parts. Which body parts do they think are involved in pollination?

   Note: Some bees may be damaged and may not have all their parts. Have students share if their bees are not complete.

3. Students not making bee sticks should follow the directions listed below:

   ■ Tell students to observe the bees with a hand lens.

   ■ Then they should read *The Bee and the* Brassica: *Interdependence* on pg. 40 in Lesson 7 of the Student Activity Book (pg. 70 in the Teacher's Guide) and any other selections on pollination in supplemental books available.

   ■ Students should devise and construct their pollination-prevention barriers so that they will be ready as soon as the first buds open.

4. When everyone has finished, collect all of the unused supplies and return them to the distribution station. Find a place to store the bee sticks. Discard the used glue cups.

Final Activities

1. Project the overhead transparency of "The Worker Bee's Body" and allow the students time to observe it, perhaps even during cleanup.

2. Initiate an observation exercise by asking the students to describe the parts of the bee. Add that they may say what they think the part is used for.

   Use the overhead projection of the bee to help the students identify:

   ■ the three main body parts: head, thorax (or midsection), and abdomen

   ■ the two large faceted eyes and the three small eyes

■ the two antennae used for touching, tasting, hearing, and smelling

■ the four wings

■ the six jointed legs

3. Assign the background reading on *Bees*, which is on pg. 35 of the Student Activity Book (pg. 57 of the Teacher's Guide).

4. Tell the students that the buds on their plants will be opening into yellow blossoms very soon. Challenge them to use the magnifier to make some observations of the buds and flowers on their own. For background information, refer them to pg. 33 in the Student Activity Book, which shows an illustration of the *Brassica* blossom with the parts labeled. The same picture is reproduced in Figure 6-2.

Figure 6-2

The Brassica *flower*

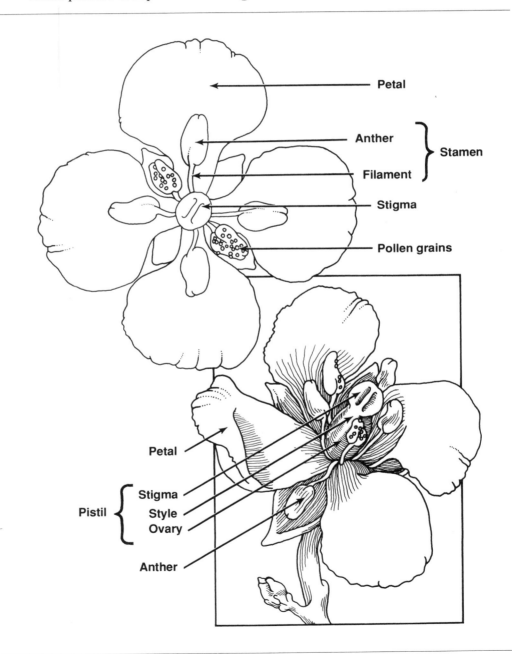

5. Remind the students experimenting with the pollination variable that they should be ready to give a brief description of their experimental plan to the class during the next lesson. They also will need to have their pollination barriers ready to put into place before the blossoms open.

6. Assign the background reading *The Bee and the* Brassica: *Interdependence* on pg. 40 in the Student Activity Book (pg. 70 in the Teacher's Guide) for the next lesson.

7. This is an important time for students to record their observations. You could remind them to include information on the size, number, and color of the leaves and buds, as well as data about their plant's height.

Extensions

1. Bees can be fascinating and are an excellent subject for library research; many outstanding trade books have been written about them (see the **Bibliography**, **Appendix B**, pg. 129). Here are some ideas for research:

    ■ A more detailed study of each of the three types of honeybees (queen, drone, and worker).

    ■ A more detailed study of the bee's anatomy and how it is specialized to do a particular job. For example, how is the queen specially suited for producing eggs, the drone for mating, and, most interesting of all, the worker for its many different jobs at different stages of its life?

    ■ The life cycle of the bee, from egg to larva to pupa to specialized adult.

    ■ How the bee perceives the world—how it sees, hears, tastes, smells, and touches.

    ■ How bees make honey.

    ■ How bees communicate by dancing.

    ■ Why bees are important to people.

2. Much can be learned about anatomy through making models. Urge the students to make anatomically correct models of honeybees and *Brassica* flowers based on their research and on their own observations. Encourage them to make the models from recycled materials. If possible, see the Science and Technology for Children unit, *Plant Growth and Development*, Lessons 13 and 14, for ideas.

3. Challenge the students to bring in some bee-related music to play while they pollinate their plants.

4. Show a film about bees and pollination (see the **Bibliography**, **Appendix B**).

5. Weather permitting, take students on a field trip to the playground. Collect, examine, then release any insects you find there.

Figure 6-3

The bee's body

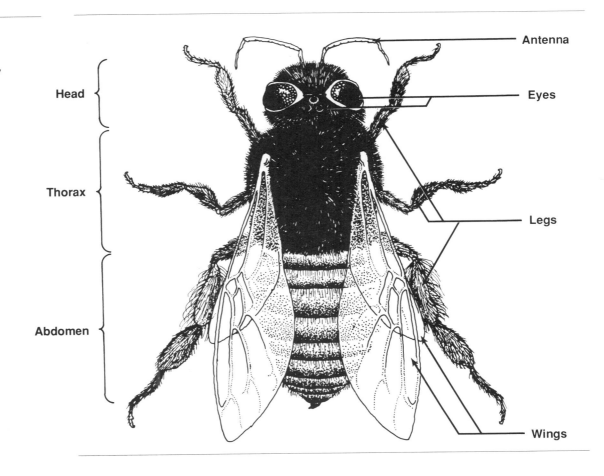

Head

Thorax

Abdomen

Antenna

Eyes

Legs

Wings

How to Make a Bee Stick **Activity Sheet 4**

NAME: _____

DATE: _____

☐ 1. Check off each item on the supply list to be sure that you have everything you need before beginning.

     ___ 1 tray   ___ 1 cup of glue (for table)

     ___ 1 dried bee   ___ 1 hand lens

     ___ 1 toothpick   ___ 1 cup (for table)

☐ 2. Observe the bee with the hand lens. Turn the bee over. Find the place where the legs are attached.

☐ 3. Put a very small drop of glue on one end of the toothpick.

☐ 4. Glue the side where the legs are to the toothpick. Make sure the head is at the end.

☐ 5. Let the glue dry for a few minutes. Be careful that the bee does not slip down the stick.

☐ 6. Now, take the time to observe the bee closely. How many of its body parts can you find? Check off each one.

Note: Some bees may be damaged and not have all of their parts. Ask a classmate to share if your bee is not complete.

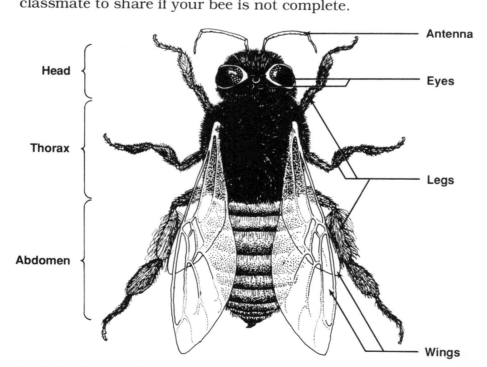

☐ 7. Push the bee stick into the bottom of an upside-down paper cup. There is one cup for each table. Leave the cup at your table. Your teacher will store the cups.

☐ 8. Place all the supplies neatly back on the tray in the center of your table.

Pollination and Interdependence

Overview

Under ideal conditions, from about Day 12 to Day 18, the blossoms will open, and students will use the bee sticks they made in Lesson 6 to **cross-pollinate** their plants. Due to delayed or slow growth, some of the experimental plants may not be ready for pollination until Day 18, 19, or later. Also, in this lesson, students who are experimenting with the pollination variable will take measures to prevent pollination from occurring.

Through readings and discussions, all students will learn more about the complexities of the interdependent relationship of the bee and the *Brassica*.

Objectives

■ Students learn more about pollination by using their bee sticks to carry it out.

■ Students experimenting with the pollination variable devise and put in place barriers to prevent pollination.

■ Students learn more about the interdependent relationship between bees and flowering plants.

Background

In nature, there are many examples of interdependent relationships, where each partner is dependent on the other. For example, there is the mutually beneficial association of cattle and the cattle egret. The cattle are useful to the birds because they provide food—ticks that have a parasitic relationship with the cattle. The birds are useful to the cattle because they free the cattle from the ticks. Between bees and flowering plants, the interdependent relationship is extremely complex.

To us, the flower is a thing of beauty that delights our senses with color and perfume. But in nature, the flower serves a specialized reproductive function; it produces seeds. In the *Brassica* flower, the male parts (six **stamens**, each composed of the **filament** and the pollen-producing **anther**) encircle the female part (one **pistil**, composed of the **stigma**, the **style**, and the **ovary**). In the **ovary** are many **ovules** that become seeds, after they are fertilized. See pg. 61 for an illustration of the parts of a flower.

For Wisconsin Fast Plants™, the pollen that helps fertilize the ovules does not come from the same flower or from a flower on the same plant. The pollen comes from a flower from another *Brassica* plant. One *Brassica* plant's own pollen cannot reach its own flower's ovules—this is prevented by a chemical reaction against the pollen. Instead, *Brassica* blossoms are **cross-pollinated**; they receive pollen from another plant, helping to ensure diversity among the *Brassica.*

For some plants, winds carry pollen from one plant to another. *Brassica's* yellow pollen grains are so heavy and sticky that they cannot be transported by the wind. Instead, pollen is moved from plant to plant by a remarkably well-adapted **pollen vector**, or pollinator—the worker bee.

To the worker bee, the *Brassica* flower represents food—nectar and pollen. Attracted to the flower by its bright color, the bee lands on a petal and thrusts her head deeply into the flower, sucking sweet nectar up through a strawlike tongue. As the bee does this, her body also brushes past the flower's **anthers**, and her hairs trap pollen. The bee also brushes against the **stigma**, which is sticky. The sticky stigma picks up pollen that was trapped in the bee's hairs from other flowers. Unwittingly, while foraging for food, the bee also has cross-pollinated the plants. The illustration on pg. 68 shows a bee pollinating a *Brassica* plant.

Before the bee flies back to the hive, she uses brushes on her midlegs to collect excess pollen from her head and thorax. Then she places the pollen in "pollen baskets" on her hind legs for transport back to the hive. There, the pollen is consumed by the bees, providing protein, fats, vitamins, and minerals.

The nectar gathered from the plant is stored by the bee in her **crop**, a nondigesting honey stomach. When the crop is full, the bee regurgitates the nectar in it into a storage cell in the hive. The nectar is the bee's source of carbohydrates. Through evaporation and the action of enzymes, the nectar becomes honey.

Bee and flower: each has benefited from the relationship. Each has provided the other with vital necessities. Each depends on the other to survive.

Materials

For each student
   1   student notebook

For every two students
   1   hand lens

For each student pollinating
   1   bee stick (made in Lesson 6)
        Plants with open flowers

For each four-member team not pollinating
        Pollination prevention barriers (made in Lesson 6)

For the class
   1   overhead transparency, "How Cross-pollination Takes Place"
        (see pg. 139)
   1   overhead transparency, "Parts of the *Brassica* Flower" (see pg. 140)
   1   overhead projector and screen

Preparation

1. Be sure to read the information in the **Background** section.

2. Set up the overhead projector and screen.

Procedure

1. Ask the nonpollinators to explain their experimental plan to the class. Have them discuss how they have kept all the variables under control. Also, students may want to show the pollination barriers they constructed to prevent accidental pollination from occurring. Suggest that they ask their classmates for comments or suggestions. Students may want to predict the outcome of this experiment. Then nonpollinators should join a teammate who is pollinating so that they can observe the process and be ready to discuss it.

2. Direct the teams that are pollinating to pick up their plants, bee sticks, and hand lenses at the distribution station. Tell them that pollination must be repeated for 5 days.

3. Tell students to cross-pollinate every blossom that is open today. This means that they must transfer pollen from the blossom on one plant to the blossom on another plant by rotating the bee gently against the anther and stigma of one flower and then against the anther and stigma of the other flower.

4. Stop the pollination process periodically and ask the students to observe with their hand lenses. Tell them to look for:

    ■ pollen on the anthers and on the stigmas

    ■ pollen trapped in the bee's hairs

5. After they have finished pollinating, students should clean up. Project the transparency, "How Cross-pollination Takes Place," on pg. 139, and allow the students to observe it for a short time, perhaps even during the cleanup period.

6. Initiate a discussion on pollination. The point of the discussion is for the students to come to the conclusion that both the bee and the blossom benefit from their relationship.

    In the discussion, ask the students what attracts the bee to the flower in real life (color and scent). Ask what the bee gets from the flower (two kinds of food: nectar and pollen). Point out that the bee must squeeze between the anthers and the stigma to reach the nectar with her strawlike tongue. Students also should have noticed the yellow pollen grains caught in the hairs of the bee's body.

    To help the students see what the flower gets from the bee, look at the transparency, "Parts of the *Brassica* Flower" (also Figure 6-2, pg. 61). It shows both the male and the female parts of the flower. Point out the female parts: the pistil with the sticky stigma at the end. Then point out the male parts: the anthers on stalks called filaments. The anthers produce the pollen.

    Ask: "How does the male part (or the pollen) of one plant get to the female part (or the stigma) of another plant?" (Through the activity of bees, of course!)

Figure 7-1

How cross-pollination takes place

Final Activities

1.   Remind the students that they must pollinate every day for the next 5 school days.

2.   Assign the reading selection *The Bee and the* Brassica: *Interdependence* on pg. 40 of the Student Activity Book. It is reproduced for you at the end of the lesson.

3.   Ask the students to predict how plants might change after pollination.

     Note: For Your Eyes Only! Let the students discover this on their own. After pollination and fertilization, the petals will fade to white, wither, and drop off, and the pistil will elongate and swell. The pistil is now a mature ovary, or a seed pod, that contains fertilized ovules, the developing seeds.

4.   Although there are no formal lessons between now and when the students harvest and thresh their crop sometime after Day 42 (a period of about three and one-half weeks), it is important that the students continue to make careful observations and to keep complete records of the growth and development of both the control and the experimental

plants during this time. It would be helpful if you could schedule the observation and recording activities for a specific time each day and give the students frequent reminders of the importance of continuing to count, measure, sketch, and write on a daily basis.

The students' record should include plant height in centimeters; number of leaves, buds, flowers, and seed pods; changes in color or texture; and the dates when important events occurred. Mention that a notation of "no change" might turn out to be just as important later as one that does indicate change.

Extensions

1. There are many library research possibilities on the topics of pollination and interdependence. Ask the students to find out more about:

    ■   other agents of pollination such as butterflies, wasps, mice, and bats

    ■   other interdependent relationships: the clown fish and the sea anemone, the horse mackerel and the Portuguese man-of-war jellyfish, the aphid and the ant, the plover and the crocodile, the shark and the remora fish.

2. Ask if anyone has an insect collection that he or she would like to bring in and share with the class.

Check the levels in the watering tanks and refill them, if necessary, through Day 35.

On Day 35, remove the plants from the water mat in order to allow the seeds to dry and ripen in preparation for harvesting and threshing (after Day 42).

After pollinating for the last time, tell the students to pinch off any unopened buds.

**Reading
Selection**

The Bee and the *Brassica:* Interdependence

The bee and the *Brassica* plant depend on each other to survive. Each takes something from the other, and each provides something to the other. You might say that they have a real partnership.

What is the relationship between the bee and the *Brassica* plant? The bee helps the plant produce seeds so that a whole new generation of plants can grow, and the plant provides the bee with food. It all starts with the fact that a *Brassica* flower has both male and female parts. One of the male parts, the anthers, produces pollen, which looks like fine yellow powder. This pollen must travel to one of the female parts, the stigma, in order for pollination, fertilization, and the production of new seeds to take place.

For the *Brassica* plants, it is very important that the pollen from one flower be transferred to the stigma of another flower on another plant. Cross-pollination is what you call the transfer of pollen from one flower on one plant to the stigma of another flower on another plant. The pollen then helps fertilize another female part—the ovules—and they become seeds.

But how does the pollen reach the other flower so that seeds can be produced? This is where the bee comes in. The bee is attracted to the flower's bright color and sweet perfume. And the flower has much to give to the bee: two kinds of food—nectar and pollen.

Picture this: the bee is dipping her head deeply into the blossom to reach the nectar deep inside. She sucks the nectar up with her strawlike tongue. As she does, her body, covered with feathery hairs, rubs up against the anthers. The hairs trap some of the pollen. When the bee flies to the next plant, some of the pollen brushes off onto the next flower's stigma.

Now the worker bee has done several jobs at once. She has collected two kinds of food from the flower, and she has carried pollen from one plant to another so that new seeds can form. Soon these seeds will grow into new plants with flowers, completing the life cycle of the *Brassica*.

Harvesting and Threshing the Seeds

Overview

The Wisconsin Fast Plants™ are nearing the end of their life cycle. If they were taken off the water at least 5 days ago, their seeds are now ripe, dry, and ready for harvest. After harvesting the crop and separating (threshing) the seeds, the students will be eager to make some final tallies. Most of the teams will find that the number and size of the pods and the number of seeds were affected by their experiment; there will be tangible evidence of the effect of their experiments on the life cycle of the *Brassica* plant.

Objectives

- Students observe that their plants have completed their life cycle, from seed to seed.

- Students observe their plants in the last stage of their lives.

- Students harvest and thresh their "crop" and collect their final data.

Background

It is reassuring to children to realize that there is a predictable order to many forms of life. Witnessing the sequential stages in the life of the *Brassica* is such an experience.

In this final stage of the life cycle, the plants are brown, brittle, and dry. Expect to hear comments such as, "I hate it that my plant died." It is important to recognize that the students have become very involved with their plants. It is equally important to point out that, although these are legitimate feelings, they are feelings and not observations. Ask if this is the end: "Is the *Brassica* dead and gone forever?" Surely an optimist in the group will protest: "No! Remember the seeds! You can plant the seeds and grow *Brassicas* again."

This is also the moment to draw some parallels between the stages of the *Brassica* plant's life and the stages of human life. The similarities are many: the seed and the fetus both develop inside a specialized structure and are nourished by the parent, both plants and humans experience an adolescent growth spurt, both develop sexual parts, both can reproduce, both eventually die.

Materials

For each student
1   student notebook
1   tray

1   envelope for storing seeds
1   quad of dried plants
1   **Activity Sheet 5, Data Record: Seeds and Pods**

Preparation

1. Distribute materials.

2. Make copies of **Activity Sheet 5**. (You may want to use this sheet only as a model, and have students set up data recording sheets in their notebooks.)

Procedure

1. Have the students break into their teams, retrieve their plants, and spend a few minutes observing the plants in their dried-out condition. The students should notice changes in color and texture. Ask each team specifically to compare their control plants with their experimental plants.

2. Pose the question: "What *new* data could you gather from these plants today?" The students will realize that the experimental plants differ from the control plants in important ways that they have not yet recorded. These differences might include the number of pods per plant, the size of each pod, and the number of seeds per pod.

3. Distribute **Activity Sheet 5** and preview it with the class or use it as a model to show the students how to set up a similar recording page in their notebooks. Then ask: "How should we collect these new data?" Accept all reasonable plans. Ensuing discussion should bring out the importance of dividing up the job and then counting carefully, double-checking the count, reporting the count honestly, and recording the results in an organized way. Be sure that everyone understands the kind of data to be collected before they begin to harvest and thresh.

4. To harvest, instruct the students either to snap the pods off with their fingers or to cut them off with scissors. To thresh the seeds, show the students how to roll each pod gently between their hands over the tray.

5. Have the students make all of the counts and measurements called for on the activity sheet. Remind them to double-check before recording a final number.

6. All of the seeds can then be placed in envelopes for storage. The envelopes should be labeled with team names, date of harvest, contents, and totals. They also should include the question the team's experiment was trying to answer and whether their plants were "experimentals" or "controls." If you plan to keep the seeds over a period of months, store them in the refrigerator in an airtight container. These seeds will be used for the germination experiments in Lessons 12 and 13.

Final Activities

1. Ask the students to discuss this topic: "Did the new data help answer your experimental question? How? Is it convincing evidence? Why or why not?"

2. Encourage the students to begin reviewing all of their data. They should now begin to plan how to organize it and to interpret what it means.

You probably will want to do a partial cleanup after this lesson.

Be sure to remember to store the seeds if you plan to do the germination experiments (Lessons 12 and 13).

Reserve at least four quads with potting mix, wicks, and fertilizer pellets if you plan to conduct the tropism experiments (Lessons 14 and 15). The remainder of the quads can be emptied and put away.

Leave the lighting system and at least one water tank in place for the tropism experiments.

Extensions

1. Wisconsin Fast Plants™ seeds do not require a dormancy period; that is, they can be planted immediately after harvesting. Challenge the students to think of new experiments that they might conduct using their seeds.

2. If you decide to conduct germination experiments in Lessons 12 and 13, tell the students that they will be able to use their own seeds. Encourage the students to do some research on germination in preparation for these new experiments.

Data Record: Seeds and Pods   **Activity Sheet 5**

NAME: _____

DATE: _____

Plant Number	Number of Pods	Number of Seeds	Length of Pods	Other Observations
1				
2				
3				
4				

Organizing and Analyzing the Data from the Experiment: Part 1

Overview

It is time for the students to reflect on the results of their team's experiment. In this lesson, they review their original question, reread their journal of daily observations, and isolate all of the quantitative data they have accumulated. The students then compare and contrast their data with those of their teammates. They make decisions on how the team's combined data can be organized best: in a graph, table, chart, drawing, or diagram. In the next lesson, they will follow through and represent their data graphically in preparation for communicating it to others.

Objectives

- Students review their team's experiment as a whole.

- Students begin to organize and analyze all of the data collected in their team's experiment.

- The team decides the best way to organize and communicate its findings.

Background

Most students will find data analysis challenging. They may need considerable guidance and reassurance. Because of this, the organizing and analyzing activities of this unit are spread over two lessons, but you may find it necessary to assign some of these activities as homework, too.

During the experiment, most of the students' data probably were recorded on tables. The students also produced a graph showing the heights to which their plants grew. These quantitative data are what the students will rely on to bring their experiments to a logical conclusion. As they stand now, these data are "raw data." They need to be processed, combined, interpreted, and communicated.

One effective way to conduct this lesson is for the students to think alone, then to compare data with a partner. Together, the partners contrast their data with the data from the rest of the team. In other words, there should be a discussion between the two like members of each team (the two students who grew control plants or the two who grew experimental plants) so that they can make comparisons. When the team gets together, they will be contrasting the findings of the two different pairs.

Materials

For each student

1   student notebook containing the daily journal of observations and the data sheets

1   **Activity Sheet 6, Data Record: How Did Each of the Variables Affect the Plants?**

For the class

1   overhead transparency entitled "How Did Each of the Variables Affect the Plants?" (see **Appendix C**, pg. 141)

1   overhead projector and screen

Preparation

1.   Obtain and set up the overhead projector and screen.

2.   Use the black-line master on pg. 141 of **Appendix C** to make an overhead transparency for use in **Procedure** No. 6.

Procedure

Note: This first activity could be assigned as homework.

1.   Allow the students (as individuals) time to reflect on their experience with the experiment. They may use **Activity Sheet 6** to organize their data. Tell them that their tasks are to:

■   be clear about what their team's experimental question is

■   reread their journal of daily observations, keeping the experimental question in mind

■   look for observations that might give clues to the answer

■   reread their data sheets, looking for trends

2.   The next step is for each student to meet with the person on the experimental team who grew his or her plants the same way and to discuss and jot down comparisons of their observations and data. (The two members who grew the control plants meet to compare the data on their two sets of normal plants, and the two members of the team who grew the experimental plants meet to compare the data on their two sets of experimental plants.) Stress that each pair is looking for likenesses in the life cycles, trends that are similar, or growth curves that are alike in some way.

3.   You may want to guide students by suggesting that they consider two key topics: a) the timing of events in the life cycle and b) measurements or quantities.

Questions that students could ask themselves on timing include:

■   When did the cotyledons (seed leaves) emerge?

■   When did the true leaves emerge?

■   When did the buds appear?

■   When did the flowers open?

Questions that students could ask themselves regarding measurements and quantities include:

■   How many of the original seeds germinated?

■   How many leaves did each plant have?

- How many flowers did each plant have?

- How many seed pods did each plant produce?

- How many seeds did each plant produce?

- How tall was each plant at different stages in its life?

4. Comparisons are not always clear-cut. It is often difficult to notice patterns. To guide the students, suggest that they look for things that appear often in the data. Ask questions relating to the frequency of events; for example, "How many of the experimental plants turned yellow on Day 12? Did all of the control plants have true leaves by Day 6?"

    Suggest, too, that the students find the average of their combined data. If all went well, the two students who grew the controls should have eight plants to work with. The experimental plants may not have fared as well (depending on how harsh the experimental conditions were) and the students who grew them may have fewer survivors to report on; this is important data, too.

5. Next, the team meets as a whole to contrast the data on the control plants with the data on the experimental plants. The point is to discover what difference changing one variable made, if any. The students should jot down their findings.

    Explain that some teams will find very obvious differences between the two sets of plants, others will find only slight differences, and some will find no differences at all. Each of these is a valid finding. Sometimes it is very important to realize that changing one particular variable has no great effect. Negative findings, or no significant differences between the experimental plant and the control plant, are just as important as positive ones. Be prepared for the fact that the students will have difficulty accepting this.

    Stress that the teams are looking for evidence of what effect the experiment had on the plants. Have them do the same kinds of comparisons as they did in Step 2 above.

6. The final task for the teams today is to decide how to organize their combined data so that they can communicate their results clearly to others. An excellent approach is to work toward displaying the data graphically. This is the time to use the overhead transparency entitled "How Did Each of the Variables Affect the Plants?" It will help you to illustrate the relationship between each of the variables manipulated and several important features of plants.

7. Guide students toward discovering for themselves the way in which the manipulation of variables has affected the plant. The list below offers suggestions of how to get students started.

    - First, ask for a volunteer to state one conclusion of his or her team's experiment. The volunteer might say, "More fertilizer makes plants develop more pods."

    - Then ask how the team came to that conclusion. Insist that they cite specific data. The students might answer, "The control plants got three fertilizer pellets each and produced an average of four pods per plant. But the experimental plants got six fertilizer pellets each and produced an average of six pods each."

    - Summarize by noting that the team members have pointed out an important relationship between the amount of fertilizer a plant gets and the number of pods it produces. Ask the students to think of a way that the data can be shown visually: by a graph, chart, table, or diagram.

Final Activities

Ask the students to think about the best ways to display the data collected in their team's experiment. In the next lesson, they will be asked to represent it visually.

Extensions

Sharpen your students' awareness of data displays in our environment. Ask students to be on the lookout for interesting graphs, tables, charts, and diagrams. They could hunt in newspapers, magazines, and advertising fliers. Post samples on a bulletin board to share with the class.

Data Record: How Did Each of the Variables Affect the Plants?   Activity Sheet 6

NAME: _____

DATE: _____

The variable I tested was: _____

	Plant 1	Plant 2	Plant 3	Plant 4
1. Height				
2. Number of leaves				
3. Number of buds				
4. Number of flowers				
5. Number of pods				
6. Number of seeds				
7. Day seeds germinated				
8. Day first true leaves developed				
9. Day first buds appeared				
10. Day first flower opened				
11. Day first petals fell				
12. Day first pods appeared				

Organizing and Analyzing the Data from the Experiment: Part 2

Overview

Today, the students pull together their data and make decisions about how to organize and represent it visually. Then they carry out their decisions, keeping in mind the goal—accurate representation of the data.

The students also write one or two sentences about the conclusion(s) they have reached based on their data. They will feel a great sense of satisfaction: they conducted a major experiment and can tell others what it means.

Objectives

■ Students organize and analyze their team's data.

■ Students represent their team's data graphically.

■ Students draw conclusions from their data.

Background

There are many ways to approach the tasks of organizing and displaying data. The object is to give a clear, concise, and accurate picture of the data—the experiment at a glance. You can then encourage the students to be as original and creative as possible as long as they accomplish this goal. Some will prefer the straightforward unadorned approach; others will unleash their artistic talents to decorate and embellish. Both are fine as long as the data is not tampered with or obscured.

Materials

For each student
   1   student notebook

For each four-member team
      Paper for graphs, charts, diagrams, and tables

Preparation

At this point, it might help to give the class a pep talk about team effort. The message is that it is important for everyone to contribute to the analysis and display of the data.

Procedure

1. Review the students' ideas of different approaches to organizing and displaying data. Be certain that students understand that there are various ways of organizing and displaying data to prove a point and that they may use any one of them.

2. Give the students time and the materials to design and construct their team's graphic data displays. As you circulate around the room, check for these trouble spots:

   ■ Have the students divided the team's work load equitably? Does everyone on the team have a part to play?

   ■ Are the students transcribing data accurately from their "raw data" collection to the new display?

   ■ Does the data display illustrate a comparison of the control group and the experimental group of plants?

   Note: After you are satisfied that the students have made a good start on the task, you may want to assign completion of the data displays as homework.

3. Now comes a crucial step: drawing a conclusion from the data. Ask each team to use one or two sentences to state the answer to their experimental question, if they can. Remind them that they should be able to point to the data that support their conclusion.

   Some teams may find that they cannot answer their experimental question or that they are not sure of their answer. This is a golden teaching moment! This happens in science laboratories all over the world. What is to be done? First, give the team plenty of credit for recognizing that they have a problem. Then try to pinpoint the possible reasons they could not answer their experimental question. They could include anything from a subtle flaw in the original plan to dumping the plants on the floor accidentally. Urge the team to report what the problems were and how they would do things differently the next time in order to get a better result.

Final Activities

Set the date for the Scientific Miniconference (see Lesson 11 for details of how to plan and run the conference).

To generate enthusiasm for the conference, tell the students that they will be interacting the way that real scientists interact. After all, what good are discoveries unless they are shared with the rest of the world?

Tell them that each team will give a presentation of their experiment. The only requirements are that:

■ the original experimental plan must be explained

■ the "raw data" for the whole team (notebooks, observations, data sheets) must be available as evidence

■ the data display must be exhibited

■ the conclusion should be stated in one or two sentences

After these requirements have been fulfilled, the students can be as creative as they wish. Encourage them to expand their participation in different directions. For example, some may wish to dress up for their presentation, add props, or even videotape the event. Others may play the role of television or newspaper reporters and interview the scientists about their work. These interviews could be published later in a school newspaper. Or the students may decide on a panel discussion format or a debate and invite an audience of parents or another class. In any case, the point is for the student scientists to communicate their results and to have fun doing it. Lesson 11 gives more complete information on how to share the data.

Figure 10-1

Displaying and communicating the data

The Scientific Conference: Communicating the Experiment and Its Results

Overview

In this lesson, students finish planning the presentation of their experiments and the results. The form and the forum of the presentation can be as simple or as elaborate as you want to make it. In its simplest form, the presentation could be an unadorned in-class oral report of the results of the experiments, given jointly by the members of each team. At its most elaborate, it could be an event, such as a scientific miniconference, complete with invited guests, cameras, roving reporters, and a *Brassica* Banquet! It is up to you and your students.

This lesson is an effective evaluation of student learning up to this point.

Objectives

■ Students plan an event to communicate the results of their experiments.

■ The class has fun and shares all they have experienced and learned together so far.

■ The teacher evaluates students' accomplishments.

Background

In organizing, formatting, and presenting their projects, the students will be exercising many different language skills and gaining valuable practice in making sense of their work in order to communicate it. This is practice that will be valuable to them in all that they do. It is what scientists do and, besides, it is fun! Just make sure that the students include the four basic requirements described below in the **Procedure** section.

Materials

For each student

      Materials needed for the various presentations

      Models of Presentation Options, pg. 52 of the Student Activity Book
        (pg. 86 of the Teacher's Guide)

Procedure

1. Each team needs to understand that no matter what form their presentation takes, it must accomplish these four objectives:

   ■ *The presentation must include an explanation of the team's experimental plan.* Suggest that the students revisit their planning boards from Lesson 2 and make them part of their exhibit. They can use the boards (enlarged, perhaps) to talk about their plan, how they arrived at it, and how it worked.

   ■ *All of the "raw data" from the whole team must be available as evidence.* Team members should organize their own notebooks, sketches, and data sheets so that they can point to the supporting evidence for their conclusion(s). They do not need to recopy or "neaten up" their original data in any way because this is their personal record of the experiment. However, they do need to have all of the data at hand and be familiar enough with it to use it to prove a point.

   ■ *The selected combined team data must be displayed graphically.* These are the graphs, tables, charts, or diagrams that the team has made by combining the individual data. The displays should show graphically how the experimental group of plants compared with the control group of plants.

   ■ *The conclusion(s) should be stated very briefly.* In one or two sentences, the students should be able to tell their conclusion(s). Of course, they have to be able to point to the evidence in their data that supports the conclusion(s).

   These are the basic requirements. Encourage the students to add to them in any way that suits their talents: board games, computer programs, crossword puzzles, find-a-word games, riddles, poems, a dramatization, or an illustrated timeline.

   MODELS OF PRESENTATION OPTIONS

   1. The *Science Fair* type of presentation

      This classical time-honored style of presenting a project has much to recommend it. It is clear, orderly, and concise, and can be worked out easily as a team project. The illustration on pg. 87 shows one of the common formats.

   2. The *Debate*

      Another possibility is to hold a debate. Many good debatable questions undoubtedly have arisen in the course of the experiments, such as: "Is more fertilizer better? Which is more important in determining a good seed harvest: light or fertilizer?"

   3. The *Panel Discussion*

      Perhaps a panel discussion of the effects on the plant of changing one variable would work well for your class. The panel could explore one question thoroughly, such as: "What are the effects of not pollinating the plant?"

   4. The *Interview*

      A "reporter" could interview a team of students about their project and then publish an article in the school newspaper or the interview could be handled as if it were for radio or television.

Figure 11-1

Presenting data in a science-fair format

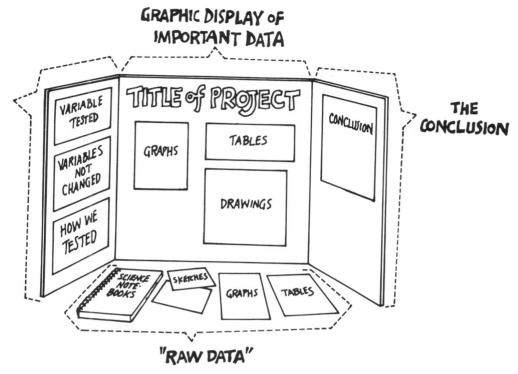

GRAPHIC DISPLAY OF IMPORTANT DATA

THE PLAN

THE CONCLUSION

VARIABLE TESTED · VARIABLES NOT CHANGED · HOW WE TESTED

TITLE of PROJECT · GRAPHS · TABLES · DRAWINGS · CONCLUSION

SCIENCE NOTEBOOKS · SKETCHES · GRAPHS · TABLES

"RAW DATA"

PLANNING AHEAD

Once you have decided what kind of conference to have, it is time to think about the details. Here is a list. Note the tasks that you and your class need to accomplish for your type of conference and check them off as you do them.

- Get whatever help you need: An extra pair of hands, someone to make phone calls, or a brainstorming partner. This is a big job, so try to recruit other adults to help.

- Invitations: Invite your guests. They might be another class, parents, the principal.

- Publicity: Notify the school newspaper, send notes home, hang posters. Attract attention!

- Refreshments: Children always remember an event that includes food. Why not feature some healthy, crunchy crucifers? (Wisconsin Fast Plants™ belong to the Crucifer family of plants; see the illustration on pg. 88.)

- Props: Each team needs to prepare its own props to enhance its presentation. These might include models of the bee and the *Brassica* plant, audiovisual paraphernalia (such as an overhead transparency and projector, a tape recorder and microphone, a video camera), or "official" identification badges.

- Furniture: Rearrange and borrow as necessary. If the students are planning a panel discussion, they will need tables and chairs set up in front of the group. If you have invited an audience, make seating arrangements. Provide the students with tables on which to display their projects.

■ Costumes: Some children may assume new roles for the conference and simple costumes make it more fun. Dress-up attire for the scientists would be appropriate. (No wild-haired, lab-coated, mad scientists, please!) Reporters may choose to impersonate their favorite newscaster and dress accordingly.

Figure 11-2

Members of the Crucifer family, relatives of Wisconsin Fast Plants

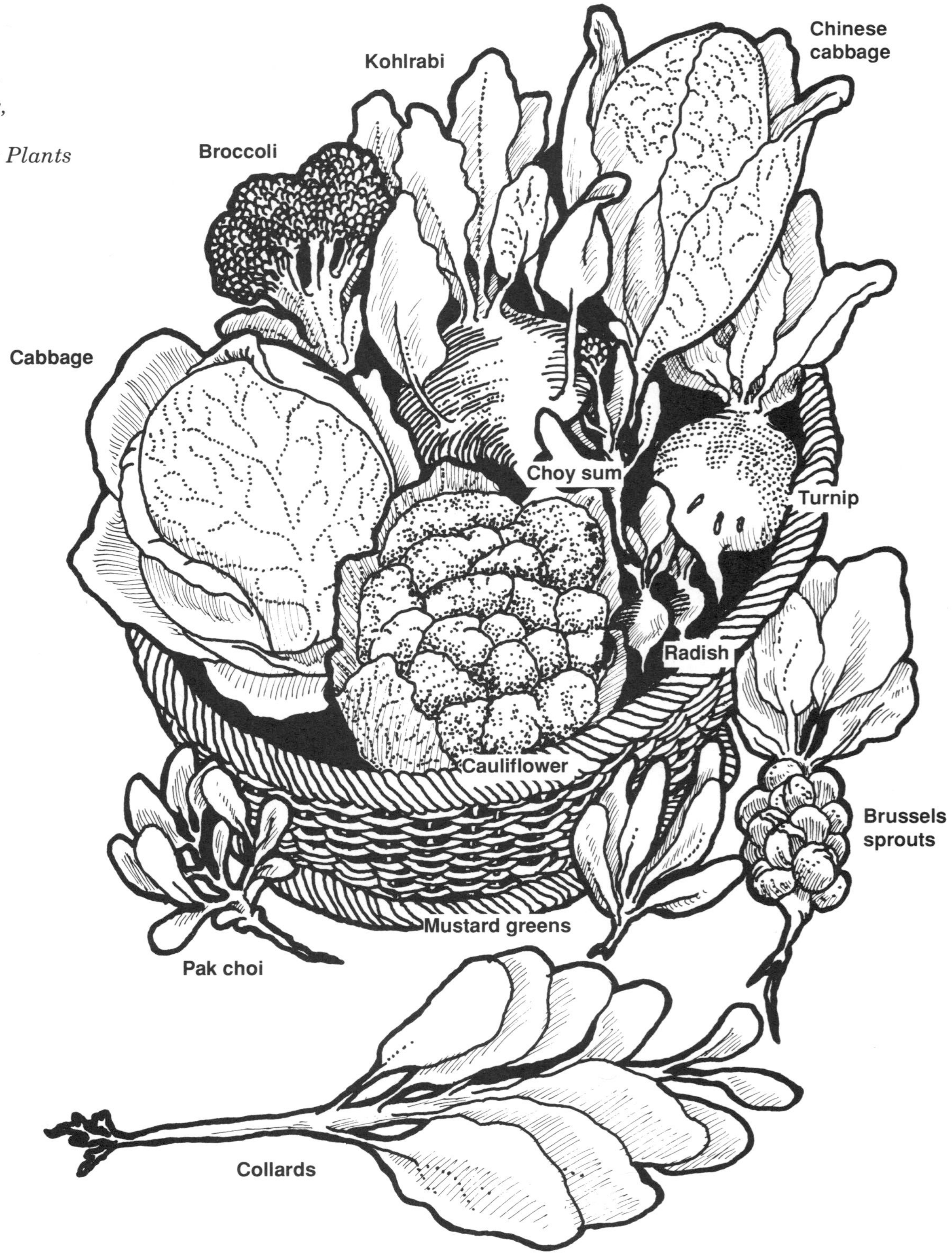

Extensions

To bring the team experiments to a close, the class could participate in one or more of the following:

- A *Brassica* Banquet: Introduce the students to the edible family of plants related to Wisconsin Fast Plants. Send a letter home asking for help in researching easy recipes that include *Brassicas* as ingredients. Then ask for volunteers to cook up a few of the delicacies or, more simply, cut up the raw crucifer vegetables and provide a savory dip.

- A filmstrip: "Around the World with Brassica" is a color filmstrip produced by the University of Wisconsin showing the economic importance of *Brassicas* in various parts of the world. It features Dr. Paul Williams, the developer of Wisconsin Fast Plants, in exotic settings with *Brassicas* all over the globe.

- A reflective writing: Ask the students to write about what they have learned from the unit. They may use their notebooks to jog their memory. Students may write a letter to a parent or friend who did not do the experiment, explaining everything that happened, or they may write a story featuring the student-researcher as the hero or heroine.

- An evaluation: Ask the students to think about what they have learned about how to do an experiment. Have their ideas about how scientists work changed? Could they plan an experiment on their own now?

Ask the students to bring in small amounts of interesting nontoxic liquids to use in the germination experiments in Lesson 12. Suggest fruit juices, soft drinks, dishwashing detergent, oil, vinegar, saltwater, and mouthwash.

Evaluation

A culminating activity such as this requires students to synthesize all that they have learned so far. Therefore, it is a valuable opportunity to evaluate. Here are some suggestions:

- How well did the team explain their experimental plan? Were they able to communicate their plan clearly?

- Was the data well organized?

- Did students prepare a graphic display of their data? Did the display show how the experimental plants compared with the control plants?

- Did students draw conclusions based on the data?

- Did students work together co-operatively in their teams to present their findings?

- Did the presentation have a spark of originality? Was there an attempt to be creative?

Planning and Setting Up Germination Experiments

Overview

This lesson and the next take advantage of the rapid life cycle of Wisconsin Fast Plants™. Students begin the cycle once again, with seeds harvested from their first set of plants. First, students individually plan and set up a seed germination experiment using seeds harvested in Lesson 8. Then in Lesson 13, students analyze the data from their experiments and draw some conclusions.

This lesson is a good opportunity for the students to review the elements of a soundly based experiment: forming a hypothesis, devising an experiment to test the hypothesis, and setting up an experiment, including the controls.

The experiment should run for 4 or 5 days, so try to begin it on a Monday or Tuesday. The next lesson should take place 5 or 6 school days from now.

These experiments are an opportunity for you to make individual assessments of what the students have learned about the experimental method.

Objectives

- Students review techniques and reinforce skills learned in planning experiments.

- Students use these skills to set up a controlled experiment in seed germination.

- The teacher evaluates how well the students can plan and set up a controlled experiment.

Background

Germination, otherwise known as sprouting, is the beginning of growth of a new plant from a seed. A seed is the ripened ovule of a flowering plant containing the embryo and **cotyledon**, or food supply, packaged inside a protective seed coat. Seeds remain **dormant**, or temporarily inactive, until conditions are suitable for the first stage of plant growth—germination. The parts of a bean seed are shown in Figure 12-1.

Figure 12-1

Parts of a bean seed

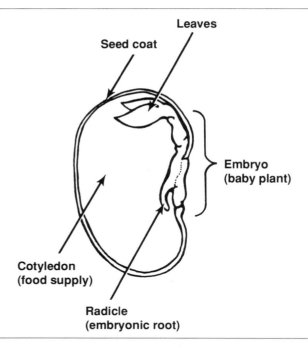

Several physical and chemical factors determine how quickly a seed germinates. The most important factors are temperature, moisture, light, oxygen, and genetic makeup. Some of these factors are potentially appropriate variables for experiments.

The first sign that the process of germination has begun is that the seed swells from having taken in water. Then the seed coat splits open. The first part to emerge from the seed is the embryonic root, called the **radicle**. It grows quickly downward, putting out fine root hairs that absorb minerals and water. Then the embryonic stem, or **hypocotyl**, pushes upward, pulling the seed leaves, or cotyledons, with it above the soil line. In Wisconsin Fast Plants™, all of this happens in 3 short days! Figure 12-2 shows these stages of growth.

Figure 12-2

How a seed germinates

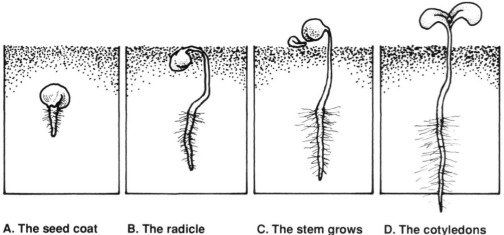

A. The seed coat splits and the embryonic root, or radicle, emerges.

B. The radicle grows downward and develops root hairs.

C. The stem grows upward and pulls the cotyledons above the soil. The seed coat falls off.

D. The cotyledons open.

Materials

For each student

1   student notebook

1   **Activity Sheet 7, Planning the Germination Experiment**

1   **Activity Sheet 8, Observations and Data Collection
     for Germination Experiment**

1   sheet of transparency film

1   paper towel

2   small resealable plastic bags

4   *Brassica* seeds (harvested in Lesson 8)

1   toothpick

For each four-member team

1   cup of water (or other liquid provided by the students)

1   dropper

For the class

1   Planning Board

    Blank index cards cut to 1½" x 5"

    Staplers

    Scissors

3   thermometers

Preparation

1.   Set up the distribution center for easy collection of materials.

2.   Try to borrow extra staplers and scissors.

Procedure

1.   Open a discussion on germination to find out what the students know about the subject already. You will also want to establish the variables for this experiment.

   Ask, "What is germination?" (The sprouting of a seed, the beginning of seed growth.) Then ask, "What time of year do seeds germinate in nature?" (Spring.) Discuss the fact that spring is the time of year when conditions of light, moisture, and temperature are favorable for plant growth. Light, moisture, and temperature also make good variables to test in a germination experiment.

2.   Display the planning board. Students have used this technique several times before, and they now should be accustomed to the pattern of thinking that it represents. Using the same procedures as before, ask the students to list the variables that it might be important to consider in a germination experiment.

3.   Write these variables on the blank cards and attach them to the planning board under "Variables we will not change." Then transfer one of the variables to the left-hand column, "Variable we will test."

4.   Discuss this variable and the kinds of testable questions you might ask about it. The planning board and variables are shown in Figure 12-3.

Figure 12-3

Planning board for the germination experiment

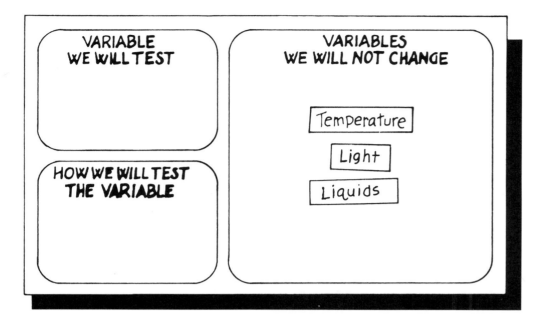

5. Below are some examples of germination topics that students have investigated successfully:

    Variable to be tested: LIGHT

    Questions:

    ■ Do seeds germinate faster under lights or in the dark?

    ■ Do seeds germinate faster in a different color light? (The student can test for this by putting a colored sheet of transparency film over the experimental germination chamber.)

    Variable to be tested: MOISTURE

    Questions:

    ■ Will seeds germinate in liquids other than water?

    Note: Students love this one. They delight in dousing their seeds with everything from ammonia to beer. Be prepared to set limits on what you can live with for 3 or 4 days.

    Variable to be tested: TEMPERATURE

    Questions:

    ■ Will seeds sprout in the refrigerator?

    ■ Will seeds sprout on the radiator?

    Note: You may find that several students want to investigate the same variable in the same way. This is fine; in fact, this is best. Scientists often try to replicate each other's experiments. And the students will collect more data on the same variable, which makes for a more convincing set of evidence.

6. Distribute **Activity Sheet 7, Planning the Germination Experiment**. Preview the sheet with the students, then allow them time to work on it to plan their individual projects. You may want to use this sheet as an evaluation tool to find out how well the students have mastered the technique of planning an experiment. (See the **Evaluation** section at the end of this lesson.)

7. Either distribute **Activity Sheet 8, Observations and Data Collection for Germination Experiments**, or use it as a model to show the students how to set up their notebooks.

8. Direct the students to the distribution station to pick up the supplies for putting together their two germination chambers. Tell them to follow the directions for assembling and labeling the chambers (one experimental, one control) on pg. 59 in their Student Activity Book. These directions are reproduced for you on pg. 97.

9. Encourage independent, creative work:

    ■ Allow the students to set up their experiments in any reasonable location. These might include tacking the plastic bags to the classroom bulletin board, hanging them from the warm cafeteria ceiling, setting them on a dark library shelf, or even snuggling them under a pillow at home.

    ■ Also, tell the students that the data charts are only suggestions of how to collect data; they are free to make improvements.

10. Ask for the students' ideas on where to keep the control germination chambers that they have set up. In order for this to be a fair test, all of the control chambers should be kept together in a location that gets light, is reasonably warm (60°F or 15°C or above), and is monitored easily. You may want to make a corner of the bulletin board or a spare table top available.

Final Activities

1. Make it clear that the students are to continue their germination experiments independently for the next 3 or 4 days, depending on what fits into your class schedule. They are expected to make daily observations and keep records. After that time, the class will come together to share data and draw conclusions.

2. Select three individuals experimenting with different variables to give brief descriptions of their experimental plans. Ask the class to comment on the strengths of the plans, to offer alternative plans, and to predict the outcome of the experiments.

Extensions

Do germination experiments with other seeds. Students may be interested in knowing if Wisconsin Fast Plants™ are faster germinators than other seeds. Try bean, radish, or grass seeds for a quick response or collect free seeds from the playground or lunch boxes. These might include acorns and dandelions or apple and orange seeds.

Evaluation

Activity Sheet 7, Planning the Germination Experiment, may be used to evaluate how well individual students have mastered the skills involved in planning a good experiment. Check that students have done the following:

    ■ selected only one variable to test

Planning and Setting Up Germination Experiments / **95**

- identified a specific testable question involving that variable
- listed something to count and something to measure
- listed some observable characteristics of the germinating seedling to record, such as color, shape, size, and time to develop
- stated a reasonable hypothesis

INSTRUCTIONS FOR HOW TO MAKE GERMINATION CHAMBERS

You will make two germination chambers, one for your control seeds and one for your experimental seeds. Follow these instruction step by step.

1.   Cut the sheet of transparency film into four pieces the same size.

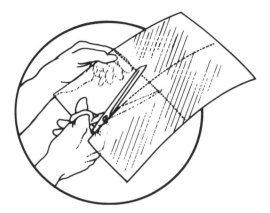

2.   Cut the paper towel into two pieces the same size as the transparency film.

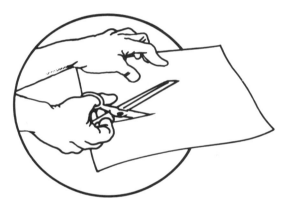

3.   Use a pencil (ink might run) to label your paper towels. Include your name and today's date. Label one towel "control," and label its two seed compartments A and B. Label the other towel "experimental" and mark its two seed compartments C and D.

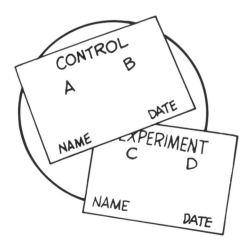

4. Put each labeled paper towel between two pieces of transparency film. Staple the towel and the transparency film together as shown. Don't get carried away; seven staples are plenty.

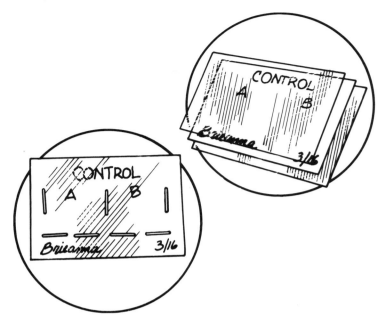

5. Use the toothpick to place one seed in each of the four compartments you labeled A, B, C, and D.

6. Put each germination chamber into a separate plastic bag and use a dropper to add water (or the liquid of your choice if this is the variable you have chosen to experiment with). Soak the paper towel thoroughly, then stop. Do not leave more than a drop or two of extra liquid standing in the bottom of the plastic bag.

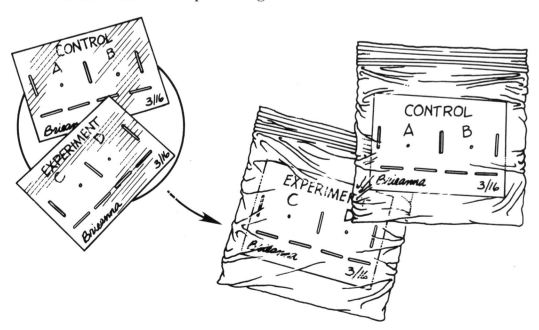

7 Close the plastic bags and place them in the locations of your choice.

8. Be sure to record the starting time of your experiment in your notebook.

Planning the Germination Experiment   **Activity Sheet 7**

NAME: _____

DATE: _____

1. The one variable I will test is: _____

2. The question I will try to answer about that variable is: _____
    _____
    _____

3. How I will test that variable: _____
    _____
    _____
    _____

4. What I will measure: _____
    _____
    _____
    _____

5. What I will count: _____
    _____
    _____
    _____

6. What I will observe: _____
    _____
    _____

7. How I will record the data: _____
    _____
    _____
    _____

8. My hypothesis is (this is what I think will happen to the seeds): _____
    _____
    _____
    _____
    _____
    _____

Observations and Data Collection for Germination Experiment   Activity Sheet 8

NAME: _____

DATE: _____

The variable I am testing is: _____

I started my experiment on _____ (date) at _____ (am/pm)

Changes in the Seed	Control Seeds		Experimental Seeds	
	Time and Date	Observations	Time and Date	Observations
Seed swells	A.		C.	
	B.		D.	
Seed coat cracks	A.		C.	
	B.		D.	
Radicle emerges	A.		C.	
	B.		D.	
Cotyledons appear	A.		C.	
	B.		D.	

What Did We Find Out about Germination?

Overview

Today, the students end their individual germination experiments. They analyze their data and draw conclusions from them. They also pool the class data.

Objectives

- Students gain further experience analyzing data and drawing conclusions from it.
- Students learn more about germination.

Background

Analyzing data may still be a difficult task for your students, but because they have had several opportunities in this unit to practice the skill, it is getting easier.

One good way to conduct this lesson is to hold a class discussion first, then to break the students into groups that investigated the same variable. Give the groups specific goals to meet in analyzing their data. Then have them report back to the class on their collective findings.

Materials

For each student
  1 student notebook
  1 **Activity Sheet 8, Observations and Data Collection for Germination Experiment** (from Lesson 12)
  2 germination chambers

For the class
1 or 2 sheets of newsprint and markers
       OR
1 or 2 overhead transparencies and markers
    1 overhead projector and screen

Preparation

1. Obtain the materials needed. On one sheet of newsprint or transparency film, write: "What did we learn about how seed germination is affected by...?" Then write the names of the variables that the students tested—light, moisture, and temperature—as subdivisions.

2. Assign the students to small discussion groups based on the variable that they investigated. (For example, place all of the students who experimented with light into one group.) If possible, divide all of the students who experimented with moisture into smaller groups according to the liquids that they investigated. This will permit more accurate comparison of results.

Procedure

1. Ask the students to spend time reviewing their own data first. They are to reflect upon their own question, decide if they have answered it, and be able to point to evidence (data) to support their answer.

2. Have the students break into their small discussion groups. Let them know that you appreciate how difficult it is to analyze data and to come to conclusions based on them. Their goal is to try to pool their group data, organize it, make sense of it, and use it to answer their original question. Make yourself available to help when needed, but urge them to look for their own conclusions.

3. Make a point of saying that there is no way to fail at this task today. Even if the students feel that their experiment has "failed" for some reason, there is something to be learned from that failure. In science, we see mistakes as learning opportunities.

4. Provide students with some leading questions to get them started. Examples are:

   ■ What happened when...? (You tried to germinate seeds in the refrigerator, for example.)

   ■ How does that compare with what happened to the control group of seeds?

   ■ How do you explain those results?

5. As the small group discussions proceed, interject more questions such as:

   ■ What part of your question can you answer today?

   ■ What evidence can you give? What data can you point to as proof?

   ■ If you repeated the experiment, do you think the results would be the same?

   ■ If different students did the same experiment, did they get the same results?

6. After the students have had sufficient time to discuss their conclusions in their small groups, bring the class back together. Tell the students that the major question before them is: "What did we find out about how seed germination is affected by light, moisture, and temperature?" You will act as recorder as the students give their conclusions. Be sure to organize the responses so that they are classified under each of the variables (light, moisture, and temperature) that the students tested.

   As students give their conclusions, ask for proof: "Why do you think that is so? What data proves it? How many seeds responded that way?"

Students will be pleased to see how much they have learned about germination and about how to make sense of data.

7. Ask the students to critique the experiment as a whole. How could it have been better? Why was it a good idea to have several students research the same topics? (They could pool their data, and reinforce each other's conclusions.)

Final Activities

In preparation for Lessons 14 and 15, the students can use their germination chambers to demonstrate tropisms. Have them tack up their control plastic bags on the bulletin board. After 1 day, have them rotate the plastic bags one quarter of a turn. Ask them to predict what direction the roots and shoots will take.

Continue rotating the plastic bags one quarter of a turn every 2 or 3 days; the plants will be going around in circles, as shown in Figure 13-1.

Figure 13-1

Geotropism in seedlings

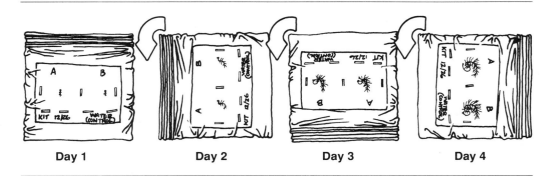

| **Day 1** | **Day 2** | **Day 3** | **Day 4** |

Plan ahead for Lessons 14 and 15. The class will need a set of at least four quads of plants that are 12 to 14 days old to conduct the experiments. You may want to assign a group of students to plant some of their surplus seeds now, in preparation for the tropism experiments in 2 weeks.

Also, assign the reading selection on *Plant Tropisms* on pg. 69 in the Student Activity Book (pg. 111 in the Teacher's Guide).

Evaluation

The students should be much more proficient now at drawing conclusions based on data. They should be able to:

- analyze their own data
- compare their data with those of other students
- use data to answer their experimental question

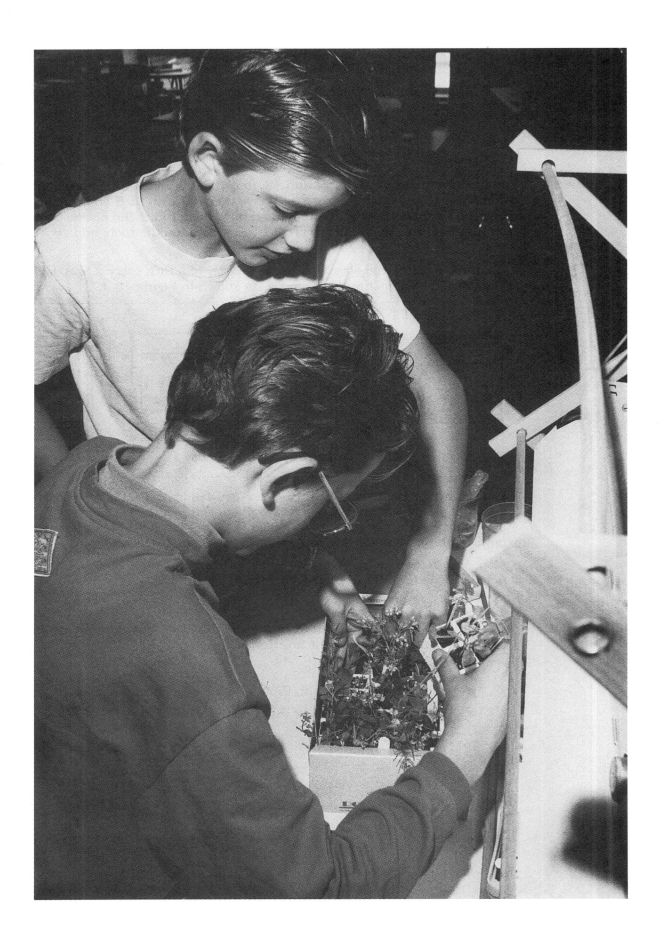

Two Tropism Experiments
(Days 12, 13, or 14)

Overview

This lesson and the next one offer students the opportunity to use all of their expertise in conducting two sets of experiments on an interesting phenomenon called **tropism.**

Objectives

- Students learn about tropisms in plants.

- Students set up experiments involving geotropism and phototropism.

- Students gain more experience in planning controlled experiments, taking measurements, and recording data.

- Students use a protractor in a real situation.

Background

Tropism is the growing or bending of a plant in response to an external source of stimulation. This lesson considers two types of tropism only: **phototropism,** which is the way a plant grows or bends in response to light, and **geotropism,** which is the way a plant grows or bends in response to gravity. These are familiar occurrences in nature that we all witness but seldom think about very much. Below are more detailed descriptions of these phenomena.

Phototropism

The geranium on the windowsill that becomes lopsided unless you remember to turn it once in a while is exhibiting phototropism. Its leaves and stems have responded positively to light by turning toward the nearest and strongest source, as shown in Figure 14-1A.

Geotropism

Generally, plants respond in two different ways to the pull of gravity: the shoots go up and the roots go down. Both of these "pulls" are shown in Figure 14-1B. To summarize, plants grow upward against gravity as well as downward toward gravity.

Figure 14-1

Examples of phototropism and geotropism

A. Phototropism—geranium turning toward the light

B. Geotropism—roots growing down and shoots growing up

This lesson assumes that students can use protractors to measure angles. If they need a review before beginning the experiment, use the exercises outlined in No. 2 in the **Extensions** section of this lesson.

Materials

For each student

1   student notebook
1   **Activity Sheet 9, Experiments in Plant Tropisms**

For the class

4   (or more) quads of *Brassica* plants 12 to 14 days old

Note: The plants respond most readily at this age. The experiments do not work as well after Day 14.

1   lightproof box
1   lighting system
16   protractors
1   overhead transparency of **Activity Sheet 9, Experiments in Plant Tropisms**
1   overhead transparency of a tree growing on a steep slope (see **Appendix C**, pg. 142)
1   planning board
Blank index cards cut to 1½" x 5"

Preparation

1.   Be sure to have available four or more quads of plants 12 to 14 days old. Suggestions for how to provide these plants were given on pg. 103 in the **Final Activities** of Lesson 13.

2.   Set up the planning board. You will use the blank cards for displaying the variables on the board as the students develop a plan for conducting the tropism experiment.

3. Devise a completely lightproof box large enough to cover one plant lying on its side as well as one plant standing upright. Possibilities include a large cardboard box or a styrofoam cooler.

4. Tell the students to read the selection entitled *Plant Tropisms* on pg. 69 of the Student Activity Book. It is reproduced on pgs. 111 and 112 of the Teacher's Guide.

5. If needed, duplicate the black line master of protractors provided in **Appendix C** on pg. 143 and cut out enough protractors for students who will be measuring during the experiment.

6. You may want to plan ahead for other lessons that will fit conveniently into the half-hour intervals between each of the measuring and recording sessions of this lesson.

Procedure

1. Display the overhead transparency of the tree growing on a steep slope (see **Appendix C**, pg. 142, and Figure 14-2). Use it to open a discussion about tropisms. Ask the students how this picture relates to what they know about tropisms (from prior knowledge and/or from reading the selection entitled *Plant Tropisms* in the Student Activity Book).

Figure 14-2

How does this tree illustrate tropisms?

Ask the students how the tree is exhibiting phototropism. Obviously, the stems and leaves are growing up, toward the light. But the case is not as clear-cut as the geranium or the sunflowers described in the reading selection. Trees do not rotate to face the sun throughout the day. At this point, you want to listen to the students' ideas.

2. The students also should discuss geotropism and how different parts of the tree respond. Although the roots meander on the surface, they respond positively to gravity and eventually go down into the soil. The trunk and branches, although a bit twisted, are basically upright; they respond negatively to gravity by growing upward.

3. Ask the students to speculate about which influence they think is stronger on a plant—light or gravity. Reassure them that there is no one right answer, but that you are interested in hearing their ideas. In fact, scientists say there is much room for research on the topic of tropisms because not enough is known about them.

4. Propose to the class that they do two experiments over the next 2 hours to find out more about how Wisconsin Fast Plants™ respond to light and to gravity. They will work toward finding an answer to the question: "Do plants grow upward in response to light?" Tell them that these two experiments will not answer the question definitively, and that scientists would perform a series of experiments in an effort to answer this question.

5. Use the planning board to discuss briefly the variables of light and position. The students have discussed light as a variable before, so this should serve merely as a confidence-building review. Position is a new idea, so the students may need more time to think about it.

6. Refer the students to pg. 66 of the Student Activity Book, which describes plans for two different experiments. Ask them to read the plans carefully. (They are summarized below.)

The Question: Do plants grow upward in response to light?

Experiment 1: In the light

How we will test: Place a quad of control plants upright under the lights. Place a quad of experimental plants on its side under the lights.

What we will measure: Measure the angle of the plant stems with a protractor and record the data on **Activity Sheet 9** every 30 minutes for 2 hours.

Experiment 2: In the dark

How we will test: Place a quad of control plants upright in a lightproof box. Place a quad of experimental plants on its side in the same lightproof box.

What we will measure: Measure the angle of the plant stems once, at the end of 2 hours. Record the data on **Activity Sheet 9**.

Note: For Your Eyes Only! Don't give it away, but under both light and dark conditions, the experimental plants will begin to bend upward at the growing tip in about 45 minutes and should have bent 90 degrees in 90 minutes. This would seem to indicate that gravity has a stronger influence on a plant than light does. Researchers are not sure about this, however.

7. Guide the students thorough an analysis of the two experimental plans. Ask them to revisit some of the criteria they applied when planning their own major experiments. Ask:

   ■ Is the question clear? Interesting?

   ■ Is the experiment doable? Will you need any special equipment?

   ■ Explain what you will measure and how. Why is it better to measure plants only once in the dark? (Double-check that the students understand how to use the protractor.)

   ■ How will the data be recorded? How often?

   ■ What kinds of observations will you make?

   ■ What do you think will happen?

8. Select students for four different jobs over the next 2 hours: students to set up the experiments, timekeepers, measurers, and recorders.

9. The students selected to set up the two experiments can do their job now. They should:

   ■ number the plants in each cell of each quad from 1 to 4

   ■ number the quads of control plants 1 and 2 and the quads of experimental plants 3 and 4.

   ■ place each of the four quads in its correct position for the experiments

10. The timekeepers can begin to keep track of the half-hour intervals and notify the squad of measurers when it is time to measure the stem angles of all of the plants. Then the recorders will post their measurements on the board for the students to copy onto **Activity Sheet 9**.

   Note: In preparation for Lesson 15 you need to record the same data as the students and transfer it to an overhead transparency that you have made from **Activity Sheet 9.**

Final Activities

1. Return all of the plants to their normal locations and positions.

2. Tell the students to clip their data sheet from both experiments into their notebooks for the next class. They will use their data to try to draw some conclusions. They also will be asked to suggest ways in which the experiment could be improved or continued in other directions.

Extensions

1. There are many possibilities for other experiments involving tropisms. Here are a few ideas:

   ■ Allow all of the plants to grow upright another day or two, then set up the same experiment a second time. Ask students if they will get the same data.

   ■ Repeat the experiment at different plant ages in order to find out more about the response of the plant's tissue at different ages.

   ■ Place a light sources below the plant. How will the stem grow?

   ■ Secure the plant and soil by stretching cut-off panty hose or cheesecloth across the top of the pot; then hang the plant upside down. In a short time, the roots will bend down and the shoot will bend up.

2. Practice with protractors. Try the exercises below.

    ■ Ask the students to determine the angle of the slope on which the tree is growing in Figure 14-1 in the Student Activity Book (see Figure 14-2 in the Teacher's Guide).

    ■ Roger uses this ramp to get to his science class each day. Measure the angle of the ramp. Is it greater or smaller than the angle of the slope on which the tree is growing? Which incline would it be harder for Roger to roll up?

Figure 14-3

How steep is the ramp?

■ When the new tree was delivered to Jenny's house, the delivery people put it on a rolling platform, then rolled it gently down a ramp from the truck to the hole. Measure the angle of this ramp.

Figure 14-4

What is the angle of this ramp?

**Reading
Selection**

Plant Tropisms

Have you ever wondered how seedlings "know" which way to grow? Why do their roots go down and their shoots go up? This seems like a simple question but, for the plant, it is a matter of life and death. It must send its roots down into the soil for minerals and water. It must send its shoot up into the light and air so that it can manufacture its food.

It is fortunate for farmers that a seedling's stems grow upward and roots grow downward no matter how they fall to the ground. Imagine the problems if that were not the case. Farmers would have to examine each seed and make sure that it was planted with the right side up. It is nearly impossible to tell which side is which on some seeds, like the tiny Wisconsin Fast Plant or the radish. Imagine how long it would take to plant an acre of seeds if you had to examine each one.

In growing straight up and down, the plant is responding to a very powerful force—gravity. This response is called "geotropism" (tropism = growing or bending in response to some force + geo = earth). What is more, parts of the plant are responding in two different ways:

- The roots are responding positively, growing toward gravity.

- The stem and leaves are responding negatively, growing up against gravity.

Figure 14-5

An example of geotropism—roots grow down and shoots grow up

Another kind of growing or bending in response to force is called "phototropism" (photo = light). It is triggered by light. Look at the following examples of phototropism.

Have you ever noticed a plant that has been sitting on a windowsill for a long time? Its stem and leaves have turned toward the light.

Figure 14-6

Why is the plant turning toward the light?

How about a field of sunflowers in the morning? (Figure 14-7A.) Then in the late afternoon (Figure 14-7B.)

Figure 14-7

A. In a field, the sunflowers face the sun in the morning

B. By late afternoon, the sunflowers look like this

Or can you imagine this? A plant will go through a maze to get to light.

Figure 14-8

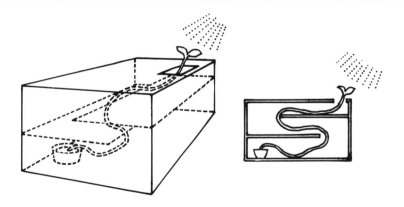

All of this leads us to some new questions. Do plants actually move? Well, yes, they do. In fact, they move continuously to survive. Some of this movement occurs during growth in stems, roots, and leaves, some of it is caused by other very slow processes that we do not easily notice.

Why must a plant move in order to survive?

Experiments in Plant Tropisms

Activity Sheet 9

NAME: _____

DATE: _____

Experiment 1: Light

	Plant No.	Angle of stem after:			
		30 minutes	60 minutes	90 minutes	120 minutes
Control Plants	1				
	2				
	3				
	4				
Experimental Plants	1				
	2				
	3				
	4				

Experiment 2: Dark

	Plant No.	Angle of stem after:			
		30 minutes	60 minutes	90 minutes	120 minutes
Control Plants	1	▬▬▬	▬▬▬	▬▬▬	
	2	▬▬▬	▬▬▬	▬▬▬	
	3	▬▬▬	▬▬▬	▬▬▬	
	4	▬▬▬	▬▬▬	▬▬▬	
Experimental Plants	1	▬▬▬	▬▬▬	▬▬▬	
	2	▬▬▬	▬▬▬	▬▬▬	
	3	▬▬▬	▬▬▬	▬▬▬	
	4	▬▬▬	▬▬▬	▬▬▬	

What Did We Find Out about Tropisms in Wisconsin Fast Plants

Overview

Today, students will use the data they generated in the two tropism experiments to try to draw some conclusions about how Wisconsin Fast Plants™ respond to light and gravity. They also will criticize the experimental designs and suggest other designs that might work as well or better. **Activity Sheet 10** included at the end of this lesson is an effective way to assess how much students have learned about tropisms.

Objectives

■ Students gain more experience analyzing data.

■ Students practice drawing conclusions based on data.

■ Students criticize the experimental plans and suggest improvements.

■ The teacher evaluates the students' understanding of tropisms.

Background

During this lesson, students analyze data from the tropisms experiments. An underlying theme of this lesson is honesty in scientific experimentation. This is a very mature theme. Depending on the maturity of your class and the level of their interest in the topic, you might incite some very lively discussions. Try peppering the lesson with some of these questions:

■ Can we trust the data? Are the measurements precise? Are they recorded accurately?

■ Do the data support the conclusion? Is it reasonable to draw that conclusion from these data? Exactly where do the data say that? Have you "jumped to a conclusion" that the data does not support?

■ Things do not always turn out the way you want them to. Should you change the data to fit your conclusion?

Materials

For each student

1　student notebook
1　completed **Activity Sheet 9, Experiments in Plant Tropisms** (from Lesson 14)

For the class

1 transparency of **Activity Sheet 9,** which you have filled in with the same data that the students have

1 overhead projector and screen

Preparation

Set up the projector and screen.

Procedure

1. Tell the students that they will use the data they collected in the two tropism experiments to try to answer the question that the experiments asked: "Do plants grow upward in response to light?" In order to answer that question, the students must scrutinize the data. Stress that every conclusion must have data to support it. The students should be able to deliver a conclusion and point directly to the data as evidence.

2. Tell the students to take out their copy of **Activity Sheet 9** from Lesson 14 and reread their data silently. They should jot down their conclusions quickly, for discussion in a few minutes. Remind them that they should be able to point to the data they are using to support their conclusions.

3. Bring the class together for a discussion of the conclusions they have gleaned from the data. Record the conclusions as students list them. Challenge each conclusion. Ask for proof.

4. Restate the original experimental question: "Do plants grow upward in response to light?" Ask the students if they can answer the question and support their answer with data. Recognize that this is difficult and ask the students if they can fathom why it is.

   The results of the two experiments are not clear-cut. There is an advantage in this: It allows the students to reexamine the original experimental plan for flaws and to suggest different approaches to answering the question. They also may see that one experiment often suggests another.

5. Because students will have difficulty coming up with an answer to the experimental question that is supported by data, a critique of the experimental plan should follow. Ask the students to criticize the way in which the two experiments were planned and to suggest other ways to find the answer. Here are some leading questions:

   ■ What worked well in the experiment? What did not work well?

   ■ Did you have enough data to answer the question? What other information would have been useful?

   ■ What new experiments could be planned to help answer the question?

   ■ How could you have improved the measurements?

   ■ How could you have collected more data from the same experiment?

   ■ How could the experiment have been controlled better? In other words, was it a fair test?

   Below are some typical responses. The responses are accurate and they indicate some good critical thinking about experimental methods.

   ■ If we used more plants, the data would be more convincing. When you say only four plants did something, it's pretty weak.

- If we measured the angles every 15 minutes, we would have a better idea of how fast the plants moved.

- We only measured the plants in the dark one time. It was for a good reason, so that they would be really in the dark the whole time, but there must be a way to measure in the dark.

- We did light and no light, but we didn't do gravity and no gravity. To be a really fair test, we need to try the experiment in a no-gravity situation. That, of course, is a little more difficult.

6. Ask the students to turn to the reading selection entitled *The Case of the Spaced-Out Bean*, which is about plants in space, on pg. 73 in their Student Activity Book. Does this help to answer the question about the effects of growing plants in a no-gravity situations? The reading selection is reproduced on pg. 118 of the Teacher's Guide.

Final Activities

Ask the students to make a brief list of new questions that they could ask about tropisms. A good title would be, "I wonder what would happen if...."

Extensions

1. Assign students to do further research into tropisms and to report their findings to the class. Two interesting examples in the plant world are the mallow for phototropism and the banyan tree for geotropism.

   The mallow is a common weed that turns its leaves to follow the sun as it moves across the sky during the course of a day. If something gets between the plant and the sun and blocks the light, the plant stops in its track. But when the sun hits it again, the plant reorients itself to the new position of the rays and resumes its motion. After dark, the mallow's leaves turn to face the east to await tomorrow's sunrise!

   The banyan tree of East India has branches that send out shoots that go down to the soil. These shoots then root and become secondary trunks.

2. There are other tropisms to explore. Assign students to research one of these:

   - chemotropism—the way a plant's roots respond to chemicals

   - thigmotropism—the way a plant's stem responds to coming into contact with an object: for example, by wrapping itself around a pole

3. As a mathematics extension, have the students translate the data from the tropism experiments onto a graph.

Evaluation

Use **Activity Sheet 10** to assess how well students understand tropisms.

**Reading
Selection**

The Case of the Spaced-Out Bean

by *Beatrice S. Smith*

Plants grown in space can't decide which way is up. At least those grown aboard the Space Shuttle Columbia couldn't.

How well plants grow in space is important because they could provide food for future space station crews. They also could keep the air fresh in the station. And they could boost the spirits of workers who are confined for months and are far removed from the good green Earth.

Scientists conducted the eight-day experiment in 1982, using pine, oat, and Chinese mung bean plants. The plants were placed aboard Columbia in two miniature gardens, and tended with care. All returned to Earth healthy enough but distorted in shape.

"More than 50 percent sent roots sprouting out of the soil instead of into it," said Dr. Joe Cowles, a University of Houston biologist. The mung bean was the worst. While in space, it twisted and turned in several directions, and never did make up its mind in which direction to grow.

On Earth, the roots of all plants grow downward into the soil toward the force of gravity. The stems grow upward toward the light. On Earth, if you tip a potted plant on its side, its roots will still eventually bend downward and its stem upward. Not so in space.

The color and size of the plants aboard Columbia were the same as those grown on Earth as control specimens. Only their shapes were different. Thus, said Dr. Cowles, the distortions obviously were caused by the absence of gravity.

Does it matter how distorted a plant's roots are?

Yes, it matters. The roots of a plant absorb water from the soil. To grow, a plant must have water. All substances that enter plant cells must do so in water. Water transports nutrients and foods up the stem to other parts of the plant. Water also keeps the cells stretched, a condition essential for them to function. And water prevents a plant from overheating.

If the roots of a plant grow upward out of the soil, they cannot properly absorb the water in the soil. A shortage of water early in the life of a plant results in retarded growth. Later in the plant's development, a water shortage may cause ripening too early and produce poor seed. If the water shortage continues, sooner or later a plant will wilt and die. Since large plants need the most food, they will die first. Smaller ones will hang on longer. But they too are doomed. No plants will grow well in space—or so it seems.

Dr. Cowles, however, is confident that in time plants can be grown successfully in space. "I think from what we have observed, it will be possible," he said. "It's not as simple as a lot of people thought. The size of the plants that can be grown may be limited. There also will have to be different ways to anchor and feed them. But I see no problem that is insurmountable."

Plans call for an American space station to be constructed sometime in the 1990's. Dr. Cowles and his co-workers have at least until then to figure out how to convince the mung bean and its companions that, if necessary, gravity is something they can manage to do without.

(Reprinted by permission of Beatrice S. Smith, © 1987)

Exercises in Plant Tropisms

Activity Sheet 10

Name: _____

Date: _____

1. What is wrong with this picture? Using what you have learned about phototropism, redraw the plant correctly.

2. What is wrong with this picture? Using what you have learned about geotropism, redraw the tree correctly.

3. Finish drawing this plant. Add 1 inch or 3 cm more of stem and six more leaves.

4. Below is an illustration of the inside of a bean seed. All seeds have two main parts: the embryo, or baby plant, and the cotyledon, or food supply. The embryo is the tiny undeveloped plant that includes all the parts of a mature plant. The food supply contains the energy the embryo needs to begin growing. All of this is enclosed in the protective seed coat.

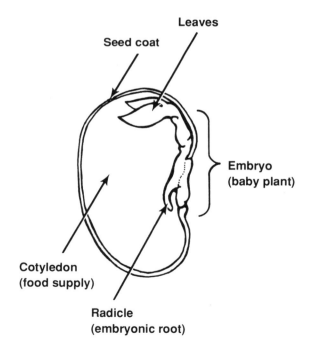

Leaves

Seed coat

Embryo (baby plant)

Cotyledon (food supply)

Radicle (embryonic root)

Here are four bean seeds planted in the ground. This is a kind of "X-ray" picture because you can see the plant embryos inside. For each seed, make a sketch showing in which direction the leaves will grow and in which direction the root will grow.

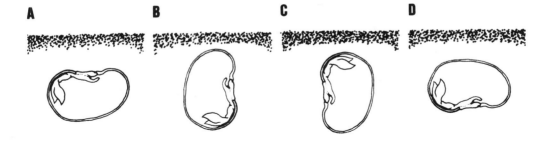

A   B   C   D

Post-Unit Assessments

Overview

- **Assessment 1** is a follow-up to the brainstorming session about plants held in Lesson 1 and a comparison of the drawing of a flowering plant from Lesson 1.

- **Assessment 2** is a rating scale that students can use to evaluate themselves.

- **Assessment 3** is a checklist of student products and process skills.

Objectives

- The students evaluate their own progress.

- The teacher evaluates the students' progress.

Materials

Because materials will vary, they are listed separately at the beginning of each assessment.

ASSESSMENT 1

A Follow-up to the Brainstorming Session about Plants Held in Lesson 1

During the brainstorming session in Lesson 1, the students developed two lists: "What we know about how to carry out an experiment" and "What we know about flowering plants." When they revisit these two lists during the assessment, the students will appreciate what they have learned through their study of the Wisconsin Fast Plant.

Materials

For each student
   1   student notebook

For the class
       The two lists saved from Lesson 1

Procedure

1. Display the two lists and analyze them with the students. Here are some points you might discuss:

   ■ Ask the students to identify statements on the lists that they now know to be true without a doubt. What experiences did they have during the project that confirmed these statements?

   ■ Ask the students to identify statements that need correction or improvement. What corrections/improvements do they want to make? Based on what?

2. Ask the students to draw and label the parts of a flowering plant.

3. When they have finished the drawing, give them an opportunity to compare this drawing to the one they did in Lesson 1.

4. Applaud students for their progress.

ASSESSMENT 2

Students' Self-Evaluation

Students use a rating scale to measure their progress.

Materials

For each student

1 **Self-Evaluation Rating Scale** (see pg. 125)

Procedure

1. Distribute a copy of the **Self-Evaluation Rating Scale** to each student, then preview it with the class. Explain that it is important to stop from time to time and think about how you are working.

2. Allow the students sufficient time to complete the rating scale, either in class or as a homework assignment.

ASSESSMENT 3

The Teacher's Record Chart of Student Progress on pg. 126 is a convenient way to assess the progress of individual students. It is designed to assist you in keeping a simple record of the work that individual students have produced as well as the skills they have developed and exhibited.

NAME: _____

DATE: _____

Rate yourself on a scale of 1 (low) to 10 (high). How well did you do each of the following activities?

Planning

1	2	3	4	5	6	7	8	9	10

Recordkeeping

1	2	3	4	5	6	7	8	9	10

Data display

1	2	3	4	5	6	7	8	9	10

Presentation

1	2	3	4	5	6	7	8	9	10

Overall use of time

1	2	3	4	5	6	7	8	9	10

Overall feelings about the experiments

1	2	3	4	5	6	7	8	9	10

Things I liked or did well: _____

_____

Things I did not like: _____

_____

Things I think I could improve: _____

_____

Next time, I would like to: _____

_____

Teacher's Record Chart of Student Progress for *Experiments with Plants*

PRODUCTS	Student																
Lesson 1: Student drawing of a flowering plant with the parts labeled (pre-test)																	
Lessons 2 & 3: Completed experiment planning board (Activity Sheets 1A and 1B)																	
Lesson 3: Completed outline of the team experiment (Activity Sheet 2)																	
Lesson 4: Record of planting (Activity Sheet 3)																	
Lesson 5: Written observations of a seedling																	
Lesson 5: Measurements of plant's height in centimeters, recorded on a graph																	
Lesson 6: Record of making a bee stick and of bee observations (Activity Sheet 4)																	
Lesson 6: Continued daily plant measurements and observations																	
Lesson 7: Continued daily plant measurements and observations																	
Lesson 8: Record of numbers of seeds and pods produced (Activity Sheet 5)																	
Lesson 9: Record of variable's effect (Activity Sheet 6)																	
Lesson 10: Data graphs																	
Lesson 10: Written conclusions																	
Lesson 12: Plan for the germination experiment (Activity Sheet 7)																	
Lesson 13: Observations and measurements of germinating control seeds and experimental seeds (Activity Sheet 8)																	
Lesson 13: Written conclusions																	
Lesson 14: Measurement of experimental plant stems and control plant stems (Activity Sheet 9)																	
Lesson 15: Paper and pencil evaluation of understanding of geotropism and phototropism (Activity Sheet 10)																	
Appendix A: Completed self-evaluation rating scale																	
Appendix A: Student drawing of a flowering plant with the parts labeled (post-test)																	

Teacher's Record Chart of Student Progress for *Experiments with Plants*

		Student														
LEARNING GOALS	Understands that a good experiment is a fair test of one variable while all other variables remain unchanged															
	Knows the requirements for the optimum growth and development of Wisconsin Fast Plants™															
	Can manipulate a planning board as a way of thinking about how to do an experiment															
	Can help design a team experiment that shows understanding of scientific methodology															
	Can maintain records of plant observations throughout an experiment, including drawings, measurements, and graphs															
	Has demonstrated the ability to plant, thin, transplant, pollinate, harvest, and thresh															
	Understands the basics of bee and flower anatomy															
	Understands the process of pollination and its importance to the production of seeds															
	Comprehends the interdependent relationship between the bee and the flowering plant															
	Can organize and analyze logically the data collected in the team's experiment															
	Can draw conclusions based on data															
	Helps present the team project, communicating the project results and conclusions															
	Can apply skills and concepts learned in the team experiment to plan and set up an individual experiment in seed germination															
	Can analyze own data on germination and draw conclusions from it															
	Can plan and set up new experiments involving phototropism and geotropism															
	Can analyze data and draw conclusions from it about tropisms															
GENERAL SKILLS	Follows directions															
	Records observations with drawings, words, and measurements															
	Works cooperatively															
	Contributes to discussions															

APPENDIX B	# Bibliography

About Bees and Pollination

Carle, Eric. *The Honeybee and the Robber: A Moving Picture Book.* New York: Putnam Publishing Group, 1981.

An entertaining book with movable parts.

Fischer-Nagel, Andreas and Heiderose Fischer-Nagel. *Life of the Honeybee.* Minneapolis: Carolrhoda Books, 1986.

Named an outstanding science trade book by the National Science Teachers Association, this book features excellent full-color pictures and a straightforward text.

Fleischman, Paul. *Joyful Noise: Poems for Two Voices.* New York: Harper & Row, 1988.

A slim volume of poetry to be read aloud by two people. The poem about honeybees is written from the points of view of the queen bee and the worker bee.

Lauber, Patricia. *From Flower to Flower.* New York: The Crown Publishing Group, 1986.

Contains remarkable black-and-white photographs of different animals in the process of pollinating plants. The text is challenging.

Parker, Nancy Winslow, and Joan Richards Wright. *Bugs.* New York: Greenwillow Books, 1987.

A lighthearted look at insects that manages to get in a lot of factual information.

About Plants and Gardening

Aliki. *Corn Is Maize: The Gift of the Indians.* New York: Harper & Row, 1976.

The story of how the Indians found and developed corn and later shared this treasure with the new settlers.

Back, Christine and Barrie Watts. *Bean and Plant*. London: A & C Black, 1984.

> Full-color photographs show how a bean plant goes from seed to seed. Easy-to-read informative text.

Bellamy, David. *The Forest*. New York: Clarkson N. Potter, Inc., 1988.
Bellamy, David. *The Roadside*. New York: Clarkson N. Potter, Inc., 1988.

> Two beautifully and cleverly illustrated books showing plant and animal inhabitants of two different ecosystems. Both books touch on the theme of interdependence.

Brown, Marc. *Your First Garden Book*. Boston and Toronto: Little, Brown & Company, 1981.

> More than twenty ideas for projects for indoor and outdoor plants; includes whimsical illustrations.

Burnie, David. *Plant*. New York: Alfred A. Knopf, 1989.

> Outstanding color photographs and lively text cover a wide range of topics: plant anatomy, pollination, seed dispersal, plant adaptations to different environments, meat eaters, and food from plants.

Cork, Barbara. *Mysteries and Marvels of Plant Life*. London: Usborne-Hayes Publishing Ltd., 1983.

> Every page is crammed with oddities about plants and unexplained marvels of nature. Full-color illustrations are both fun and informative.

Heller, Ruth. *The Reason for a Flower*. New York: Grosset & Dunlap, 1983.

> A sprinkling of text on brilliantly illustrated pages conveys a wealth of information about plants, pollination, and plant products. A favorite that makes for easy reading.

Holley, Brian. *Plants and Flowers*. Burlington, Vermont: Hayes Publishing Ltd., 1986.

> Especially good illustrations of parts of flowers.

Jaspersohn, William. *How the Forest Grew*. New York: Greenwillow Books, 1980.

> The chronicle of a hardwood forest from its beginnings 200 years ago to full maturity. Easy-to-read text and detailed pen-and-ink drawings.

Kerven, Rosalind. *The Tree in the Moon and Other Legends of Plants and Trees*. New York: Cambridge University Press, 1989.

> Stories derived from folk legends collected from all over the globe. Beautifully illustrated.

Oechsli, Helen and Kelly Oechsli. *In My Garden*. New York: Macmillan Publishing Co., 1985.

> A child's step-by-step guide to planting and maintaining a garden. Includes techniques for thinning, transplanting, weeding, and composting.

Pranis, E. and J. Hale. *Grow Lab: A Complete Guide to Gardening in the Classroom.* Burlington, Vermont: National Gardening Association, 1988.

Contains a wealth of information and practical suggestions for growing plants indoors under lights. Also rich in ideas for integrating science into the rest of the curriculum.

Schnieper, Claudia. *An Apple Tree through the Year.* Minneapolis: Carolrhoda Books, Inc., 1987.

Full-color photographs and text explaining the yearly cycle of the growth and development of fruit.

Suzuki, David. *Looking at Plants.* New York: Warner Books, 1985.

A book of simple plant projects for indoors and out.

Thompson, Ruth. *Usborne First Nature Trees.* London: Usborne-Hayes Publishing Ltd., 1980.

An amply illustrated book featuring a wide variety of trees.

Williams, Paul H., Coe M. Wiliams, and Richard P. Green. *Exploring with Wisconsin Fast Plants.* Madison, Wisconsin: University of Wisconsin, 1990.

An elementary/middle school teacher's manual written by the team that developed Wisconsin Fast Plants™. The manual is imbued with that team's spirit of enthusiasm for scientific investigation. Conveniently divided into five sections full of information on such topics as materials, ideas for subsequent investigations, extensions in language arts and games, and supplementary materials for the teacher.

Wyler, Rose. *Science Fun with Peanuts and Popcorn.* New York: Julian Messner, 1986.

Simple projects to help students explore germination, growth, and development. Includes fun activities such as tongue twisters and riddles.

About How Scientists Work

Kramer, Stephen P. *How to Think Like a Scientist: Answering Questions by the Scientific Method.* New York: Thomas Y. Crowell, 1987.

Uses engaging questions to introduce children to the process of solving problems by using the scientific method. Gives excellent and clear examples.

National Science Teachers Association. *Science Fairs and Projects.* Washington, D.C.: NSTA, 1984.

A compilation of magazine articles written by teachers to help teachers plan, develop, and exhibit projects.

Smith, Norman. *How Fast Do Your Oysters Grow?* New York: Messner, 1982.

> Describes the process of doing a science project, from selecting a topic to designing apparatus, keeping records, drawing conclusions, and reporting results.

Stwertka, Eve and Albert Stwertka. *Make It Graphic! Drawing Graphs for Science and Social Studies Projects.* New York: Messner, 1985.

> Very good examples of graphic displays of data. Shows graphs of all types.

Biography

Franchere, Ruth. *Cesar Chavez.* New York: Harper-Collins, 1970.

> Describes how Chavez organized farm workers to protest unfair treatment.

Mitchell, Barbara. *Pocketful of Goobers.* Minneapolis: Carolrhoda Books, Inc., 1986.

Moore, Eva. *The Story of George Washington Carver.* New York: Scholastic, Inc., 1971.

> Both books tell the story of how Carver overcame enormous obstacles to become an outstanding plant scientist. He is best known for inventing hundreds of uses for the peanut.

Supplemental

Computer Programs

"Botanical Gardens." Available from: Sunburst Communications, Inc., 39 Washington Ave., Pleasantville, New York 10570.

> Apple II series. This software program provides a straightforward way for students (grade 6 and up) to practice the process of conducting controlled experiments on plants. It is flexible enough to permit advanced students to design their own plants in a "genetics lab" and allows teachers to control simply the kinds of computer-generated "growing experiences" that students have.

"MECC Lunar Greenhouse" and "MECC Graph." Available from: MECC, 3490 Lexington Ave. N., St. Paul, Minnesota 55126.

> Apple II series; printer recommended. Children collect data, enter information, and generate graphs. For grades 7 to 9, but easily adapted for younger children. The greenhouse program also lets students experiment with variables.

"Project Zoo: Adventures with Charts and Graphs." Available from: National Geographic Society Educational Services, Department 89, Washington, D.C. 20036.

> Apple II series; color monitor and printer recommended, but optional. Contains 3 disks, filmstrip, audio cassette, manual, work sheets, zoo fact book, pretest, and posttests. For grades 3 to 5.

Video

"Pollination." Available from: National Geographic Society, Educational Services, Washington, D.C. 20036.

> Illustrates flower anatomy, pollination, and fertilization. Outstanding footage of ordinary as well as exotic pollinators. For grades 4 to 9. 23 minutes.

"What's Buzzin'." Available from: Cimarron Products, 3131 S. Vaughn Way, Suite 134, Aurora, Colorado 80014.

Filmstrips

"How Living Things Depend on Each Other." Available from: National Geographic Society, Educational Services, Washington, D.C. 20036.

> Designed to teach the interdependence of living things. Comes with cassette. For intermediate grades. 13 minutes.

"Insects: How They Help Us." Available from: National Geographic Society, Educational Services, Washington, D.C. 20036.

> Emphasizes the beneficial aspects of insect life. Comes with cassette. For intermediate grades. 15 minutes.

"Scientists and How They Work." Available from: National Geographic Society, Educational Services, Washington, D.C. 20036.

> Shows many different kinds of scientists at work both in the field and in the laboratory. Describes some of the methods that scientists use in their work. For grades 4 to 6. 17 minutes.

Transparencies

"Honeybee II Symbiosis with Flowering Plants." Available from: Carolina Biological, Burlington, North Carolina 27215.

> Beautifully drawn and colored, the transparencies show the honeybee anatomy in relation to the brassica plant. Advanced, but easily adapted for younger children.

"Around the World with Brassicas." Available from: University of Wisconsin-Madison, Department of Plant Pathology, 1630 Linden Drive, Madison, Wisconsin 53706.

> A collection of twenty color slides describing *Brassicas* and their economic importance as food for people and animals.

Black Line Masters

VARIABLE
WE WILL TEST

HOW WE WILL TEST
THE VARIABLE

VARIABLES
WE WILL NOT CHANGE

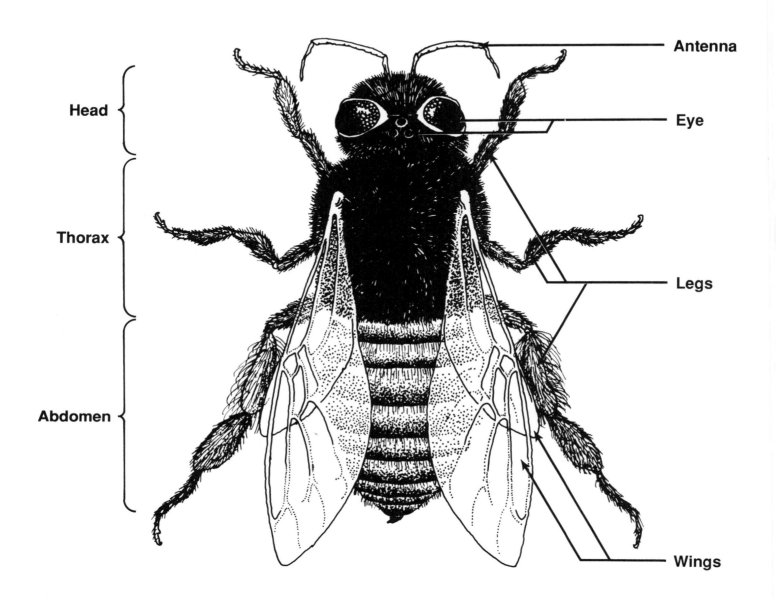

Head

Thorax

Abdomen

Antenna

Eye

Legs

Wings

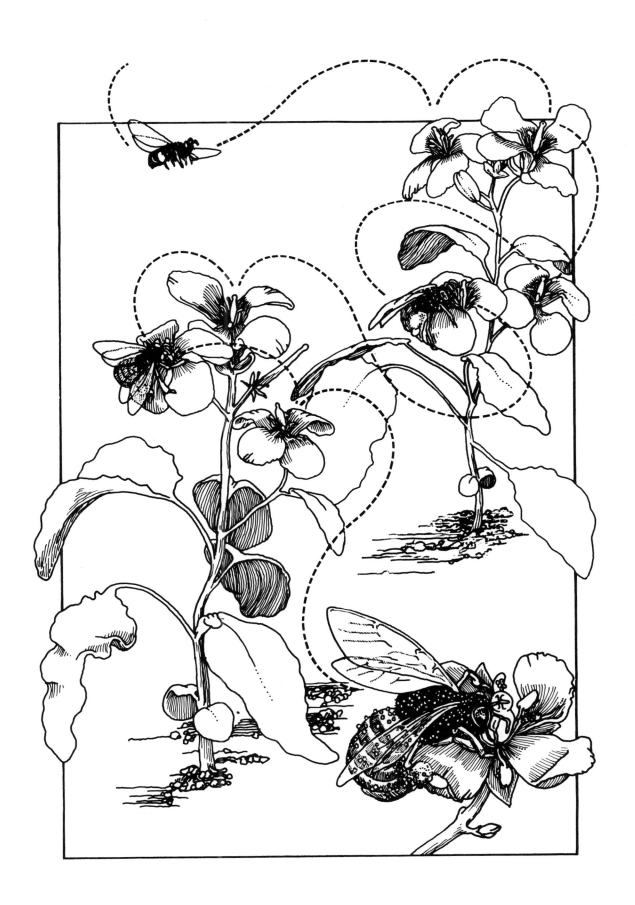

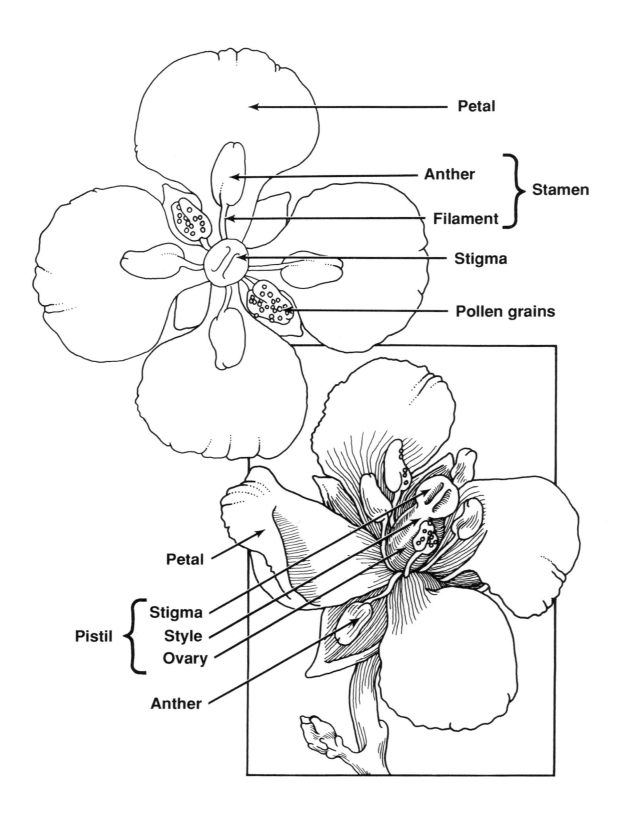

Petal

Anther   } Stamen

Filament

Stigma

Pollen grains

Petal

Pistil { Stigma

Style

Ovary

Anther

How Did Each of the Variables Affect the Plants?

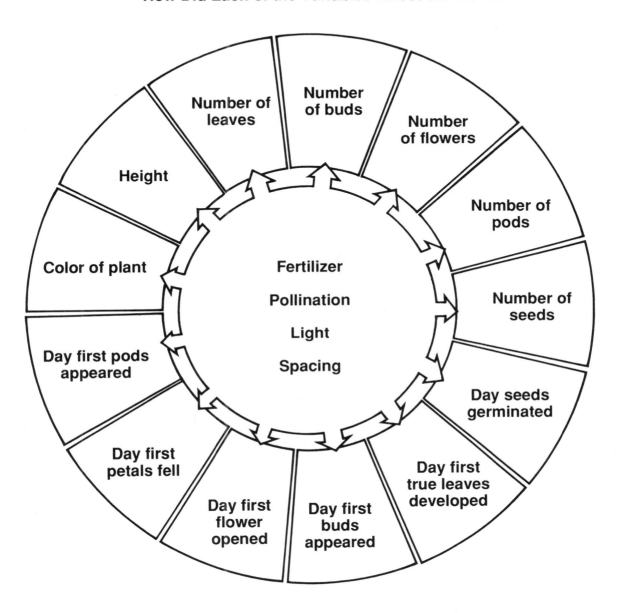

Number of leaves

Number of buds

Number of flowers

Height

Number of pods

Color of plant

Fertilizer

Pollination

Light

Spacing

Number of seeds

Day first pods appeared

Day seeds germinated

Day first petals fell

Day first true leaves developed

Day first flower opened

Day first buds appeared

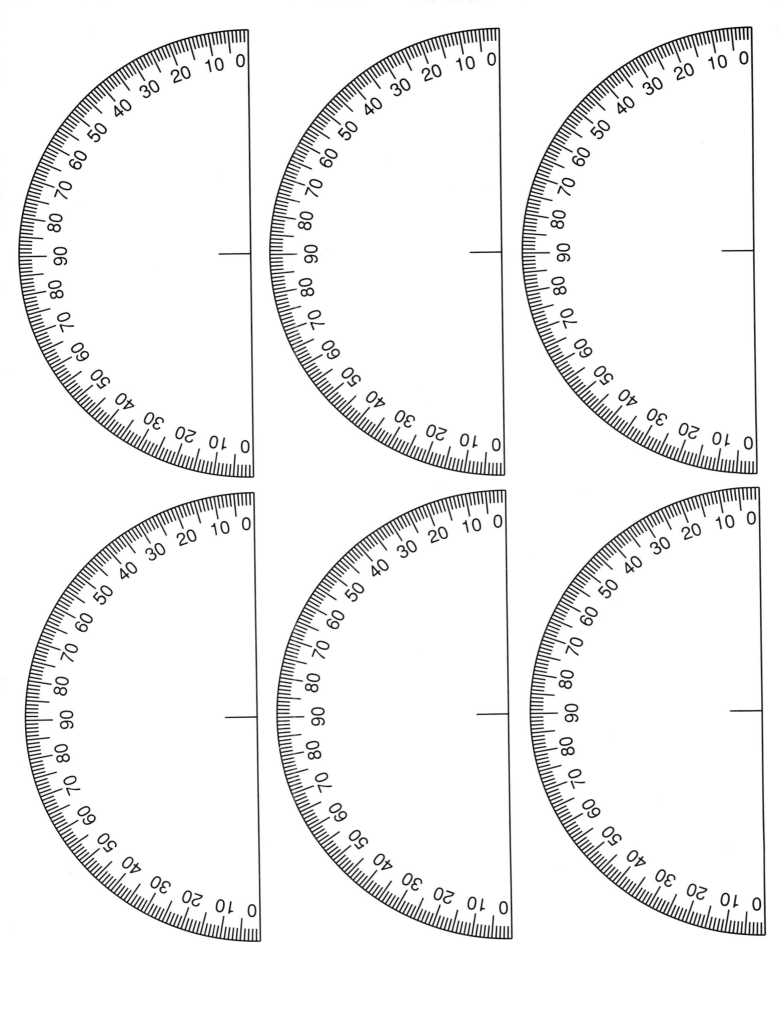

Materials Reorder Information

During the course of hands-on science activities, some of the materials are used up. The consumable materials from each Science and Technology for Children unit can be reordered as a unit refurbishment set. In addition, a unit's components can be ordered separately.

For information on refurbishing *Experiments with Plants* or purchasing additional components, please call Carolina Biological Supply Company at **800-334-5551** and ask for an STC Customer Service Representative.